The Perfectionists

ALSO BY SIMON WINCHESTER

Pacific

In Holy Terror

American Heartbeat

Their Noble Lordships

Stones of Empire *(photographer)*

Outposts

Prison Diary, Argentina

Hong Kong: Here Be Dragons

Korea: A Walk Through the Land of Miracles

Pacific Rising

Small World

Pacific Nightmare

The River at the Center of the World

The Professor and the Madman

The Fracture Zone

The Map That Changed the World

Krakatoa

The Meaning of Everything

A Crack in the Edge of the World

The Man Who Loved China

West Coast: Bering to Baja

East Coast: Arctic to Tropic

Skulls

Atlantic

The Alice Behind Wonderland

The Men Who United the States

When the Earth Shakes

When the Sky Breaks

Oxford

The Perfectionists

How Precision Engineers Created the Modern World

Simon Winchester

An Imprint of HarperCollinsPublishers

Much of the material on pages 492–97 relating to the Tohoku Tsunami of March 2011 is taken with permission from an essay by Simon Winchester in the *New York Review of Books*, November 9, 2017.

HarperCollins books may be purchased for educational, business, or sales promotional use. For information please e-mail the Special Markets Department at SPsales@harpercollins.com.

FIRST HARPERLUXE EDITION

ISBN: 978-0-06-284590-0

HarperLuxe™ is a trademark of HarperCollins Publishers.

Library of Congress Cataloging-in-Publication Data is available upon request.

18 19 20 21 22   ID/LSC   10 9 8 7 6 5 4 3 2 1

For Setsuko

And in loving memory of my father,
Bernard Austin William Winchester, 1921–2011,
a most meticulous man

These brief passages from works by the writer Lewis Mumford (1895–1990) might usefully be borne in mind while reading the pages that follow.

The cycle of the machine is now coming to an end. Man has learned much in the hard discipline and the shrewd, unflinching grasp of practical possibilities that the machine has provided in the last three centuries: but we can no more continue to live in the world of the machine than we could live successfully on the barren surface of the moon.

—*THE CULTURE OF CITIES* (1938)

We must give as much weight to the arousal of the emotions and to the expression of moral and esthetic values as we now give to science, to invention, to practical organization. One without the other is impotent.

—*VALUES FOR SURVIVAL* (1946)

Forget the damned motor car and build the cities for lovers and friends.

—*MY WORKS AND DAYS* (1979)

Contents

List of Illustrations

Unless otherwise noted, all images are in the public domain.

Prologue

The aim of science is not to open the door to infinite wisdom, but to set a limit to infinite error.

—BERTOLT BRECHT, *Life of Galileo* (1939)

We were just about to sit down to dinner when my father, a conspiratorial twinkle in his eye, said that he had something to show me. He opened his briefcase and from it drew a large and evidently very heavy wooden box.

It was a London winter evening in the mid-1950s, almost certainly wretched, with cold and yellowish smog. I was about ten years old, home from boarding school for the Christmas holidays. My father had come in from his factory in North London, brushing flecks of gray industrial sleet from the shoulders of his army officer's greatcoat. He was standing in front of the coal fire to warm himself, his pipe between his teeth. My mother was bustling about in the kitchen, and in time she carried the dishes into the dining room.

But first there was the matter of the box.

I remember the box very well, even at this remove of more than sixty years. It was about ten inches square and three deep, about the size of a biscuit tin. It was evidently an object of some quality, well worn and cared for, and made of varnished oak. My father's name and initials and style of address, B. A. W. WINCHESTER ESQ., were engraved on a brass plate on the top. Just like the much humbler pinewood case in which I kept my pencils and crayons, his box had a sliding top secured with a small brass hasp, and there was a recess to allow you to open it with a single finger.

This my father did, to reveal inside a thick lining of deep red velvet with a series of wide valleys, or grooves. Firmly secured within the grooves were a large number of highly polished pieces of metal, some of them cubes, most of them rectangles, like tiny tablets, dominoes, or billets. I could see that each had a number etched in its surface, almost all the numbers preceded by or including a decimal point—numbers such as .175 or .735 or

1.300. My father set the box down carefully and lit his pipe: the mysterious pieces, more than a hundred of them, glinted from the coal fire's flames.

He took out two of the largest pieces and laid them on the linen tablecloth. My mother, rightly suspecting that, like so many of the items my father brought home from the shop floor to show me, they would be covered with a thin film of machine oil, gave a little cry of exasperation and ran back into the kitchen. She was a fastidious Belgian lady from Ghent, a woman very much of her time, and spotless linen and lace therefore meant much to her.

My father held the metal tiles out for me to inspect. He remarked that they were made of high-carbon stainless steel, or at least another alloy, with some chromium and maybe a little tungsten to render them especially hard. They were not at all magnetic, he added, and to make his point, he pushed them toward one another on the tablecloth—leaving a telltale oil trail to further upset my mother. He was right: the metal tiles showed no inclination to bond with each other, or to be repelled. Pick them up, my father said, take one in each hand. I took one in each palm and made as if to measure them. They were cold, heavy. They had heft, and were rather beautiful in the exactness of their making.

He then took the pieces from me and promptly

placed them back on the table, one of them on top of the other. Now, he said, pick up the top one. Just the top one. And so, with one hand, I did as I was told—except that upon my picking up the topmost piece, the other one came along with it.

My father grinned. Pull them apart, he said. I grasped the lower piece and pulled. It would not budge. Harder, he said. I tried again. Nothing. No movement at all. The two rectangular steel tiles appeared to be stuck fast, as if they were glued or welded or had become one—for I could no longer see a line where one tile ended and the other began. It seemed as though one piece of steel had quite simply melted itself into the structure of the other. I tried again, and again.

By now I was perspiring from the effort, and my mother, back from the kitchen, was getting impatient, and so my father set his pipe aside and took off his jacket and began to dish out the food. The tiles were beside his water glass, symbols of my muscular impoverishment, my defeat. Could I have another try? I asked at dinner. No need, he said, and he picked them up and with a flick of his wrist simply slid one off the other, sideways. They came apart instantly, with ease and grace. I was openmouthed at something that, viewed from a schoolboy's perspective, seemed much like magic.

No magic, my father said. All six of the sides, he

explained, are just perfectly, impeccably, exactly flat. They had been machined with such precision that there were no asperities whatsoever on their surfaces that might allow air to get between and form a point of weakness. They were so perfectly flat that the molecules of their faces bonded with one another when they were joined together, and it became well-nigh impossible to break them apart from one another, though no one knows exactly why. They could only be slid apart; that was the only way. There was a word for this: *wringing*.

My father started to talk animatedly, excitedly, with a passionate intensity that I always liked. Metal tiles like these, he said, and with a very evident pride, are probably the most precise things that are ever made. They are called gauge blocks, or Jo blocks, after the man who invented them, Carl Edvard Johansson, and they are used for measuring things to the most extreme of tolerances—and the people who produce them work at the very summit of mechanical engineering. These are precious things, and I wanted you to see them, since they are so important to my life.

And with that said, he fell quiet, carefully put the gauge blocks back in their velvet-lined wooden box, finished his dinner, lit his pipe once more, and fell asleep by the fire.

My father was for all his working life a precision engineer. In the closing years of his career, he designed and made minute electric motors for the guidance systems of torpedoes. Most of this work was secret, but once in a while he would smuggle me into one of his factories and I would gaze in either admiration or puzzlement at machines that cut and notched the teeth for tiny brass gearwheels, or that polished steel spindles that seemed no thicker than a human hair, or that wound copper coils around magnets that seemed no bigger than the head of a pipe smoker's vesta.

I remember with great fondness spending time with one of my father's favored workers, an elderly man in a brown lab coat who, like my father, clasped a pipe between his teeth, leaving it unlit all the time he worked. He wore a permanently incised frown as he sat before the business end of a special lathe—German, my father said; very expensive—watching the cutting edge of a notching tool as it whirled at invisible speed, cooled by a constant stream of a cream-like oil-and-water mixture. The machine hunted and pecked at a small brass dowel, skimming as it did so microscopic coils of yellow metal from its edges as the rod was slowly rotated. I watched intently as, by some curiously magical process, an array of newly cut tiny teeth steadily appeared incised into the metal's outer margins.

The machine stopped for a moment; there was a sudden silence—and then, as I squinted into the moving mass of confusion around the workpiece, a gathering of separate and more delicate tungsten carbide tools moved into view and were promptly engaged, and the spindles began to turn and cut, such that the teeth that had so far been created were now being shaped and curved and notched and chamfered, the machine's magnifying glass showing just how the patterns of their edges evolved as they passed beneath the blades, until, with a whisper of disengagement, the spinning stopped, the dowel was sliced as a side of ham might be, the clamp was released, and out of a filter lifted from the cream-oil bath rose a dripping confection of impossibly shiny finished gearwheels, maybe twenty of them, each no more than a millimeter thick and perhaps a centimeter in diameter.

They were all flipped by an unseen lever out of the lathe and onto a tray, where they would lie ready to be slipped onto spindles and then attached in mysterious fashion to the motors that turned a fin here or varied the pitch of a screw there, with the gyroscopically ordered intention of keeping a high-explosive submarine weapon running straight and true toward its enemy target through the unpredictable movements of a cold and heaving sea.

Except that, in this case, the elderly craftsman decided that the Royal Navy could easily spare one from this fresh batch of wheels. He took a pair of steel needle-nose tweezers and picked a sample out of the creamy bath, washed it under a gush of clear water, and handed it to me with an expression of pride and triumph. He sat back, smiled broadly at a job well done, and lit a satisfying pipe. The tiny gearwheel was a gift, my father would say, a reminder of your visit. As precise a gearwheel as you'll ever see.

Just like his star employee, my father took singular pride in his profession. He regarded as profound and significant and *worthy* the business of turning shapeless slugs of hard metal into objects of beauty and utility, each of them finely turned and neatly finished and fitted for purposes of all imaginable kinds, prosaic and exotic—for as well as weaponry, my father's plants built devices that went into motorcars and heating fans and down mineshafts; motors that cut diamonds and crushed coffee beans and sat deep inside microscopes, barographs, cameras, and clocks. Not watches, he said ruefully, but table clocks and ships' chronometers and long-case grandfather clocks, where his gearwheels kept patient time to the phases of the moon and displayed it on the clock dials high up in a thousand hallways.

He would sometimes bring home pieces even more elaborate than but perhaps not quite as magical as the gauge blocks, with their ultra-flat, machined faces. He brought them primarily to amuse me, unveiling them at the dinner table, always to my mother's chagrin, as they were invariably wrapped in oily brown wax paper that marked the tablecloth. Will you put that on a piece of newspaper? she'd cry, usually in vain, as by then the piece was out, shining in the dining room lights, its wheels ready to spin, its arms ready to be cranked, its glassware (for often there was a lens or two or a small mirror attached to the device) ready to be demonstrated.

My father had a great fascination with and reverence for well-made cars, most especially those made by Rolls-Royce. This at a time, long past, when these haughty machines represented not so much the caste of their owners as the craft of their makers. My father had once been granted a tour of the assembly line in Crewe and had spent a while with the team who made the engine crankshafts. What impressed him most was that these shafts, which weighed many scores of pounds, had been finished by hand and were so finely balanced that, once set spinning on a test bench, they had no inclination to stop spinning, since no one side was even fractionally heavier than another. Had there been no such phenomenon as friction, my father said, a Phantom V's crankshaft, once set spinning, could run in perpetuity. As a result of that conversation, he had me try to design a perpetual motion machine of my own, a dream on which I wasted (given my then only very vague understanding of the first two laws of thermodynamics, and thus the impossibility of ever meeting the challenge) many hours of spare time and many hundreds of sheets of writing paper.

Though more than a half century has elapsed since those machine-happy days of my childhood, the memory still exerts a pull—and never more so than one afternoon in the spring of 2011, when I received, quite unexpectedly, an e-mail from a complete stranger in the town of Clearwater, Florida. It was headed simply "A Suggestion," and its first paragraph (of three) started without frill or demur: "Why not write a book on the History of Precision?"

My correspondent was a man named Colin Povey, whose principal career had been as a scientific glassblower.* The argument he put forward was persuasive in its simplicity: precision, he said, is an essential component of the modern world, yet is invisible, hidden in plain sight. We all know that machines have to be precise; we all recognize that items that are of importance to us (our camera, our cellphone, our computer, our bicycle, our car, our dishwasher, our ballpoint pen) have to sport components that fit together with precision and operate with near perfection; and we all probably suppose that the more precise things are, the better they are. At the same time, this phenomenon of precision, like oxygen or the English language, is something we take for granted, is largely unseen, can seldom be fully imagined, and is rarely properly discussed, at least by

---

* The few hundred members of this somewhat exclusive calling specialize in making glass instruments of great delicacy and complexity for use largely in chemical laboratories. They have a journal, *Fusion*; they hold conventions; and they have a hero, a Japanese American immigrant named Mitsugi Ohno, who, until his death in 1999 at age seventy-three, worked mainly for Kansas State University and whose collection of enormous and detailed glass models of ships and iconic American buildings remains on the campus in the town of Manhattan. Ohno is most famous for having found a way to blow a Klein bottle, a recurving vessel that, like a three-dimensional version of a Möbius strip, has only a single surface.

those of us in the laity. Yet it is always there, an essential aspect of modernity that makes the modern possible.

Yet it hasn't always been so. Precision has a beginning. Precision has a definite and probably unassailable date of birth. Precision is something that developed over time, it has grown and changed and evolved, and it has a future that is to some quite obvious and to others, puzzlingly, somewhat uncertain. Precision's existence, in other words, enjoys the trajectory of a narrative, though it might well be that the shape of that trajectory will turn out to be more a parabola than a linear excursion into the infinite. In whichever manner precision developed, though, there was a story; there was, as they say in the moviemaking world, a through line.

That, said Mr. Povey, was his understanding of the theory of the thing. Yet he also had a personal reason for suggesting the idea, and to illustrate it, he told me the following tale, which I offer here in summary, a mix of precision and concision:

Mr. Povey Sr., my correspondent's father, was a British soldier, a somewhat eccentric figure by all accounts who, among other things, classified himself as a Hindu so that he would not be obliged to attend the normally compulsory Sunday Anglican service. Not wishing to fight in the trenches, he joined the Royal Army Ordnance Corps, the body that has the responsibility of supplying weapons, ammunition, and armored vehicles to those soldiers who used such things in battle. (The RAOC's functions have since expanded, and now, less glamorously, it also runs the army's laundry and mobile baths and does the official photography.)

During training, he learned the rudiments of bomb disposal and other technical matters, excelling at the engineering aspects of the craft, and thus qualified, he was sent in 1940 to the British embassy in Washington, DC (in secret, and wearing civilian clothes, as the United States had so far not joined the war). His duties were mainly to liaise with American ammunition makers to create ordnance that would fit into British-issued weapons.

In 1942, he was given a special mission: to work out just why some American antitank ammunition was jamming, randomly, when fired from British guns. He promptly took a train to the manufacturers in Detroit and spent weeks at the factory painstakingly measuring batches of ammunition, finding, to his chagrin, that every single round fitted perfectly in the weapon for which it was destined, meeting the specifications with absolute precision. The problem, he told his superiors back in London, did not lie with the plant. So London told him to follow the ammunition all the way to where the commanders were experiencing the vexing misfires, and that was in the battlefields of the North African desert.

Mr. Povey, lugging along his giant leather case of measuring equipment, promptly lit out for the East Coast. He first traveled on a variety of ammunition trains, passing slowly across the mountains and rivers of eastern America, all the way to Philadelphia, whence the ordnance was to be shipped. Each day, he measured the shells, and found that they and their casings retained their design integrity perfectly, fitting the gun barrels just as well at each of the railway depots as they had when they left the production lines. Then he boarded the cargo ship.

It turned into something of a testing journey: the vessel broke down, was abandoned by its convoy and its destroyer escort, became frighteningly vulnerable to attack by U-boats, and was trapped in a mid-ocean storm that left all of the crew wretchedly seasick. But, as it happened, it was this deeply testing environment that allowed Mr. Povey finally to solve the puzzle.

For it turned out that the severe rocking of the ship damaged some of the shells. They were stacked in crates deep in the ship's hold. As the vessel rocked and heeled in the storm, those crates on the outer edges of the stacks, and only those, would crash into the sides of the ship. If they hit repeatedly, and if when they hit they were configured in such a way that the tip of the ammunition struck the wall of the hold, the whole of the metal projectile at the front end of each shell—the bullet, to put it simply—would be shoved backward, by perhaps no more than the tiniest fraction of an inch, into its brass cartridge case. This collision, if repeated many times, caused the cartridge case to distort, its lip to swell up, very slightly, by a near-invisible amount that was measurable only by the more sensitive of Povey's collection of micrometers and gauges.

The shells that endured this beating—and they would be randomly distributed, for once the ship had docked and the stevedores had unloaded the crates and the ammunition had been broken down and sent out to the various regiments, no one knew what order the shells would be in—would, as a result, not fit into the gun barrels out on the battlefield. There would, in consequence, be (and entirely randomly) a spate of misfires of the guns.

It was an elegant diagnosis, with a simple recommended cure: it was necessary only for the factory back in Detroit to reinforce the cardboard and wood of the ammunition crates and—presto!—the shell casings would all emerge from the ship unbruised and undistorted, and the jamming problem with the antitank rifles would be solved.

Povey telegraphed his news and his suggestion back to London, was immediately declared a hero, and then, in classic army style, was equally immediately forgotten about, in the desert, without orders, but with, as he had been away from his office in Washington for so long, a considerable amount of back pay.

Hot work in the Sahara it must have been, for at this point the story wavers a little: Mr. Povey Sr. seems to have gone on some kind of long-drawn-out desert bender. But after enjoying the sunshine for an indecent number of weeks, he decided that he did in fact need to return to America, so he bribed his way back there with bottles of Scotch whisky. It took him eleven bottles of Johnnie Walker to get from Cairo (via a temporary aerodrome in no less exotic a wartime stopover than Timbuktu) to Miami, after which it was but a short and easy hop up to Washington.

There he found dismaying news. It turned out that he had been away in Africa for so long without any communication that he had been declared missing and presumed dead. His mess privileges had been revoked, his cupboard closed, and all his clothes altered to fit a much smaller man.

It took a while for this discomfiting mess to be sorted

out, and when eventually everything was more or less back to normal, he discovered that his entire ordnance unit had been transferred to Philadelphia—to which he promptly went as well.

There he met and fell in love with the unit's American secretary. The pair got married, and Mr. Povey, never apparently practicing the Hinduism that had been engraved on his army dog tag, remained blamelessly in the United States for the rest of his days.

And, as my correspondent then wrote, with a flourish, "the lady in question was my mother, and so I exist—and I exist entirely because of precision." This is why, he then added, "you must write this book."

Before we delve too deeply into its history, two particular aspects of precision need to be addressed. First, its ubiquity in the contemporary conversation—the fact that precision is an integral, unchallenged, and seemingly essential component of our modern social, mercantile, scientific, mechanical, and intellectual landscapes. It pervades our lives entirely, comprehensively, wholly. Yet, the second thing to note—and it is a simple irony—is that most of us whose lives are peppered and larded and salted and perfumed with precision are not, when we come to think about it, entirely sure what precision is, what it means, or how it differs from similar-sounding concepts—accuracy most obviously, or its lexical kissing cousins of perfection and exactitude and of being *just right, exactly!*

Precision's omnipresence is the simplest to illustrate.

A cursory look around makes the point. Consider, for example, the magazines on your coffee table, in particular the advertising pages. In a scant few minutes you could, for instance, construct from them a rough timetable for enjoying a precision-filled day.

You would begin your morning by first using a Colgate Precision Toothbrush; if you were clever enough to keep up with Gillette's many product lines, you could enjoy less "tug and pull" on your cheek and chin by shaving with the "five precision blades" in its new Fusion5 ProShield Chill cartridge, and then tidying up your goatee and mustache with a Braun Precision Trimmer. Before the first meeting with a new acquaintance, be sure to have any former-girlfriend-related body art painlessly removed from your biceps with an advertised machine that offers patented "precision laser tattoo removal." Once thus purified and presentable, serenade your new girlfriend by playing her a tune on a Fender Precision bass guitar; maybe take her for a safe wintertime spin after fitting your car with a new set of guaranteed-in-writing Firestone Precision radial snow tires; impress her with your driving skills first out on the highway and then at the curb with adroit use of the patented Volkswagen Precision parking-assist technology; take her upstairs and listen to soft music played on a Scott Precision radio (a device that will add "laurels of magnificent dignity to those of the world-record achievements" of the Chicago-based Scott Transformer Company—not all the magazines on an average coffee table are necessarily current). Then, if the snow has eased, prepare dinner in the back garden with a Big Green Egg outdoor stove equipped with "precision temperature control"; gaze dreamily over nearby fields newly sown with Johnson Precision corn; and finally, take comfort from the knowledge that if, after the stresses of the evening, you awake hungover or unwell, you can take advantage of the precision medicine that is newly available at NewYork-Presbyterian Hospital.

It took no time at all to tease out these particular examples from one randomly selected coffee-table pile. There are all too many others. I see, for instance, that the English novelist Hilary Mantel recently described the future British queen, née Kate Middleton, as being so outwardly perfect as to appear "precision-made, machine-made." This went down well with neither royalists nor engineers, as what is perfect about the

Duchess of Cambridge, and indeed with any human being, is the very imprecision that is necessarily endowed by genes and upbringing.

Precision appears in pejorative form, as here. It is also enshrined elsewhere and everywhere in the names of products, is listed among the main qualities of the function or the form of these products, is all too often one of the names of companies that produce such products. It is also used to describe how one uses the language; how one marshals one's thoughts; how one dresses, writes by hand, ties ties, makes clothing, creates cocktails; how one carves, slices, and dices food—a sushi master is revered for the precise manner in which he shaves his *toro*—how cleverly one throws a football, applies makeup, drops bombs, solves puzzles, fires guns, paints portraits, types, wins arguments, and advances propositions.

QED, one might say. *Precisely.*

Precision is a much better word, a more apposite choice in all the examples just given, than is its closest rival, *accuracy.* "Accurate Laser Tattoo Removal" sounds not nearly as convincing or effective; a car with merely "Accurate Parking Technology" might well be assumed to bump occasional fenders with another; "Accurate Corn" sounds, at best, a little dull. And it surely would be both damning and condescending to say that you tie your tie accurately—to knot it precisely is much more suggestive of élan and style.

The word *precision*, an attractive and mildly seductive noun (made so largely by the sibilance at the beginning of its third syllable), is Latin in origin, was French in early wide usage, and was first introduced into the English lexicon early in the sixteenth century. Its initial sense, that of "an act of separation or cutting off"—think of another word for the act of trimming, *précis*—is seldom used today:* the sense employed so often these days that it has become a near cliché has to do, as the *Oxford English Dictionary* has it, "with exactness and accuracy."

In the following account, the words *precision* and *accuracy* will be employed almost but not quite interchangeably, as by common consent they mean just about the same thing, but not exactly the same thing—not precisely.

Given the particular subject of this book, it is important that the distinction be explained, because to the true practitioners of precision in engineering, the

---

* Although T. S. Eliot did employ it in his 1917 "Rhapsody on a Windy Night": "Whispering lunar incantations / Dissolve the floors of memory / And all its clear relations, / Its divisions and precisions . . ."

difference between the two words is an important one, a reminder of how it is that the English language has virtually no synonyms, that all English words are specific, fit for purpose by their often very narrow sense and meaning. *Precision* and *accuracy* have, to some users, a significant variation in sense.

The Latin derivation of the two words is suggestive of this fundamental variance. *Accuracy's* etymology has much to do with Latin words that mean "care and attention"; *precision*, for its part, originates from a cascade of ancient meanings involving separation. "Care and attention" can seem at first to have something, but only something rather little, to do with the act of slicing off. Precision, though, enjoys a rather closer association with later meanings of *minuteness* and *detail.* If you describe something with great accuracy, you describe it as closely as you possibly can to what it is, to its true value. If you describe something with great precision, you do so in the greatest possible detail, even though that detail may not necessarily be the true value of the thing being described.

You can describe the constant ratio between the diameter and the circumference of a circle, pi, with a very great degree of *precision*, as, say, 3.14159265 358979323846. Or pi can happily be expressed with *accuracy* to just seven decimal places as 3.1415927—this being strictly accurate because the last number, 7, is the mathematically acceptable way to round up a number whose true value ends (as I have just written, and noted before the gap I have placed in it) in 65.

A somewhat simpler means of explaining much the same thing is with a three-ring target for pistol shooting. Let us say you shoot six shots at the target, and all six shots hit wide of the mark, don't even graze the target—you are shooting here with neither accuracy nor precision.

Maybe your shots are all within the inner ring but are widely dispersed around the target. Here you have great accuracy, being close to the bull's-eye, but little precision, in that your shots all fall in different places on the target.

Perhaps your shots all fall between the inner and outer rings and are all very close to one another. Here you have great precision but not sufficient accuracy.

Finally, the most desired case, the drumroll result: your shots are all clustered together *and* have all hit the bull's-eye. Here you have performed ideally in that you have achieved both great accuracy and great precision.

In each of these cases, whether writing the value of pi or shooting at a target, you achieve accuracy when the accumulation of results is close to the desired value, which in these examples is either the true value of

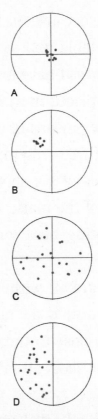

The image of a target offers an easy means of differentiating precision and accuracy. In A, the shots are close and clustered around the bull: there is both precision and accuracy. In B, there is precision, yes, but insofar as the shots miss the bull, they are inaccurate. C, with the shots widely dispersed, shows neither precision nor accuracy. And in D, with some clustering and some proximity to the bull, there is moderate accuracy and moderate precision—but very moderate.

the constant or the center of the target. Precision, by contrast, is attained when the accumulated results are similar to one another, when the shooting attempt is achieved many times with exactly the same outcome—even though that outcome may not necessarily reflect the true value of the desired end. In summary, accuracy is true to the intention; precision is true to itself.

One last definition needs to be added to this mass of confusion: the concept of tolerance. Tolerance is an especially important concept here for reasons both philosophical and organizational, the latter because it forms the simple organizing principle of this book. Because an ever-increasing desire for ever-higher precision seems to be a leitmotif of modern society, I have arranged the chapters that follow in ascending order of tolerance, with low tolerances of 0.1 and 0.01 starting the story and the absurdly, near-impossibly high tolerances to which some scientists work today—claims of measurements of differences of as little as 0.000 000 000 000 000 000 000 000 000 01 grams, *10 to the -28th grams*, have recently been made, for example—toward the end.★

★ Crucial to the making of almost anything is the matter of its measurement. In English, this usually involves the use of the near-invisible adverb *how*, with its interrogative determination of *to what extent* and *to what degree* something might be. *How long is it, how massive, how straight an edge, how curved a surface, how hard, how close the fit?* The Ancient Egyptians were the first to define such terms, with the *cubit*, the length

Yet this principle also prompts a more general philosophical question: why? Why the need for such tolerances? Does a race for the ever-increasing precision suggested by these measurements actually offer any real benefit to human society? Is there perhaps a risk that we are somehow fetishizing precision, making things to ever-more-extraordinary tolerances simply because we

---

of a pharaoh's forearm, generally agreed to be the grand old man of measurement. Thereafter other human attributes were derived by other civilizations—the length of a thumb or a foot, the distance covered by a thousand paces, the span of a day's journeying—to form the basis of measuring scales, with the inch or the pound or the grave or the catty usually fixed; while others—the Chinese unit of distance, the *li*, for example—were set to vary, depending on whether the road to be traveled was flat or uphill. Then came the French and their deliciously neat and decimally based *système métrique*, and in short order thereafter today's sedulously contrived and internationally agreed International System of Units, which is better known as SI, and which defines (and has been formally adopted by all, except for Burma, Liberia, and the United States) the seven fundamental units of length, mass, time, electric current, temperature, amount of substance, and light intensity—otherwise known as the meter, the kilogram, the second, the ampere, the kelvin, the mole, and the candela. In order not to stall the narrative pace of this history, I have placed a more detailed tour of measurement's multitudinous mysteries at the end of this book, as an afterword.

can, or because we believe we should be able to? These are questions for later, but they nonetheless prompt a need here to define tolerance, so that we know as much about this singular aspect of precision as about precision itself.

Although I have mentioned that one may be precise in the way one uses language, or accurate in the painting of a picture, most of this book will examine these properties as far as they apply to manufactured objects, and in most cases to objects that are manufactured by the machining of hard substances: metal, glass, ceramics, and so forth. Not wood, though. For while it can be tempting to look at an exquisite piece of wooden furniture or temple architecture and to admire the accuracy of the planing and the precision of the joints, the concepts of precision and accuracy can never be strictly applied to objects made of wood—because wood is flexible; it swells and contracts in unpredictable ways; it can never be truly of a fixed dimension because by its very nature it is a substance still fixed in the natural world. Whether planed or jointed, lapped or milled, or varnished to a brilliant luster, it is fundamentally inherently imprecise.

A piece of highly machined metal, however, or a lens of polished glass, an edge of fired ceramic—these can be made with true and lasting precision, and if the manufacturing process is impeccable, they can be made time and time again, each one the same, each one potentially interchangeable for any other.

Any piece of manufactured metal (or glass or ceramic) must have chemical and physical properties: it must have mass, density, a coefficient of expansion, a degree of hardness, specific heat, and so on. It must also have dimensions: length, height, and width. It must possess geometric characteristics: it must have measurable degrees of straightness, of flatness, of circularity, cylindricity, perpendicularity, symmetry, parallelism, and position—among a mesmerizing host of other qualities even more arcane and obscure.

And for all these dimensions and geometries, the piece of machined metal must have a degree of what has come to be known* as tolerance. It has to have a tolerance of some degree if it is to fit in some way in a

---

* Since its first formal definition, in 1916, as "the permissible margins of error" in machine workmanship. An 1868 British report on international coinage presaged this particular usage when it noted that as far as gold coins were concerned, "the margin for error in coining . . . known as the remedy or tolerance . . . amounts to 15 grains for the fineness, plus or minus 1/16 of a carat."

machine, whether that machine is a clock, a ballpoint pen, a jet engine, a telescope, or a guidance system for a torpedo. There is precious little point in tolerance if the machined object is simply to stand upright and alone in the middle of a desert. But to fit with another equally finely machined piece of metal, the piece in question must have an agreed or stated amount of permissible variation in its dimensions or geometry that will allow it to fit. That allowable variation is the tolerance, and the more precise the manufactured piece, the greater the tolerance that will be needed and specified.

A shoe, for instance, is invariably a thing of very low tolerance: on the one hand, a poorly made slipper may have "an agreed or stated amount of allowable variation in its dimensions" (which is the engineer's formal definition of tolerance) of half an inch, with so generous an amount of wiggle room between foot and lining as to make the notion of precision almost irrelevant. A handmade brogue shoe by Lobb of London, on the other hand (or foot), may seem to fit snugly, perfectly, precisely even, but it will still have a tolerance of maybe an eighth of an inch—and in a shoe, such a tolerance would be acceptable, and the shoe indeed worn with pride. Yet, in terms of precision engineering, it is anything but precisely made; nor is it even accurately so.*

One of the two most precise measuring instruments ever built by human agency stands in America's Pacific Northwest, far away from everything, in the arid middle of Washington State. It was built just outside the top-secret nuclear installation where the United States created the first supplies of plutonium for the bomb that destroyed Nagasaki, for decades the material at the heart of much of the nation's arsenal of atomic weapons.

The years of nuclear activity there have left an unimaginably large legacy of dangerously irradiated substances, from old fuel rods to contaminated items of clothing, which are only now, and after a loud public outcry, being remedied—or *remediated*, the term environmentalists prefer. Today, the Hanford site, as it is known, is officially the largest environmental cleanup site in the world, with decontamination bills reaching the tens of billions of dollars and the necessary remedial work likely to last until the middle of the twenty-first century.

---

* Precision-made shoe lasts, created on a machine designed by one Thomas Blanchard in Springfield, Massachusetts, in 1817, are a part of the American precision story, too, as will be explained in chapter 3.

I first passed by the site very late one night, after a long drive from Seattle. From my southbound speeding car, I could see the glimmer of lights in the far distance. Behind razor-wire security fences and warning signs and under the protection of armed guards, some eleven thousand workers are now laboring night and day to cleanse the earth and waters of the poisonous radioactivity that so dangerously suffuses it. Some suppose it is a task so vast that it may never be properly completed.

To the south of the main cleanup site, just outside the razor-wire fence but within sight of the still-standing towers of the remaining atomic piles, one of present-day science's most remarkable experiments is being conducted. It is not secret at all, is unlikely to leave a legacy of any danger whatsoever, and requires the making and employment of an array of the most precise machines and instruments that humankind has ever attempted to construct.

It is an unassuming place, easily missed. I arrived for my appointment in morning daylight, weary after the long nighttime drive. It was cold; the road was quite empty, the main turnoff unmarked. A small notice on the left pointed to a cluster of low white buildings a hundred yards off the road. "LIGO," the sign read. "WELCOME." And that was about all. Welcome to the current cathedral, it might also have said, to the worship of ultraprecision.

It has taken decades to design the scientific instruments that are secreted out in the middle of this dust-dry nowhere. "We maintain our security by our obscurity" is the motto for those who fret about the costly experiments sited there, all without a fragment of barbed wire or chain link to protect them. The tolerances of the machines at the LIGO site are almost unimaginably huge, and the consequent precision of its components is of a level and nature neither known nor achieved anywhere else on Earth.

LIGO is an observatory, the Laser Interferometer Gravitational-Wave Observatory. The purpose of this extraordinarily sensitive, complex, and costly piece of equipment is to try to detect the passage through the fabric of space-time of those brief disruptions and distortions and ripples known as gravitational waves, phenomena that in 1916 Albert Einstein predicted, as part of his general theory of relativity, should occur.

If Einstein was right, then once every so often, when huge events occur far out in deep space (the collision of a pair of black holes, for instance), the spreading fan of interstellar ripples, all moving at the speed of light, should eventually hit and pass through the Earth and, in doing so, cause the entire planet to change shape, by an infinitesimal amount and for just the briefest moment of time.

No sentient being would ever feel such a thing; and the slight squeezing would be so minute and momentary and harmless that not a trace could ever be recorded by any machine or device known—except, in theory, by LIGO. And after decades of experiments with instruments that were being ever more refined to greater and greater degrees of sensitivity, the devices now running in the high northwest desert of Washington State and down in the bayous of Louisiana, where the second such observatory has been built, have indeed brought home the bacon.

For, in September 2015, almost a century after Einstein's theory was first published, and then again on Christmas Eve that same year and then again in 2016, LIGO's instruments showed without doubt that a series of gravitational waves, arriving after billions of years of travel from the universe's outer edges, had passed by and through Earth and, for the fleeting moment of their passage, changed our planet's shape.

To detect this, the LIGO machines had to be constructed to standards of mechanical perfection that only a few years before were well-nigh inconceivable and that, before then, were neither imaginable nor even achievable. For it was not always so, this delicacy, this sensitivity, this ultraprecise manner of doing things. Precision was not always there, waiting in the shadows, needing to be found and then exploited for what its early admirers believed would be the common good. Far from it.

Precision was a concept that was invented, quite deliberately, out of a single and well-recognized historic need. It was brought into being for severely practical reasons—reasons that had much to do not with any dreamy twenty-first-century wish to confirm (or otherwise) the existence of vibrations from the collisions of distant stars. Rather, it had to do with a down-to-earth eighteenth-century realization of what was then a pressing matter of physics, and which was related to the potentially awesome power of that high-temperature form of water that since the century before had been known as and defined by the word *steam.*

Precision's birth derives from the then-imagined possibility of maybe holding and managing and directing this steam, this invisible gaseous form of boiling water, so as to create power from it, and to demand that by the employment of this power, it perform useful work for the good (perhaps, and with luck) of all humankind.

And all that, what turned out to be one of the most singular of engineering epiphanies, took place in North

Wales on a cool May day in 1776—by coincidence, within weeks of the founding of the United States of America, which would eventually make such use of the precision techniques that duly evolved.

That spring day is now generally (though not unanimously) agreed to mark the birth date for the making of the first construction possessed of a degree of real and reproducible mechanical precision—precision that was measurable, recordable, repeatable, and, in this case, created to the tolerance of one-tenth of an inch, or, as it was put at the time, of an English silver coin with a value or worth of just one shilling.

Chapter 1

(Tolerance: 0.1)

Stars, Seconds, Cylinders, and Steam

It is the mark of an instructed mind to rest assured with that degree of precision that the nature of the subject admits, and not to seek exactness when only an approximation of the truth is possible.

—ARISTOTLE (384–322 BC), *Nicomachean Ethics*

The man who by the common consent of the engineering fraternity is regarded as the father of true precision was an eighteenth-century Englishman named John Wilkinson, who was denounced sardonically as lovably mad, and especially so because of his passion for and obsession with metallic iron. He made an iron boat, worked at an iron desk, built an iron pulpit, ordered that he be buried in an iron coffin, which he kept in his workshop (and out of which he would jump to amuse his comely female visitors), and is memorialized by an iron pillar he had erected in advance of his passing in a remote village in south Lancashire.

Still, a case can also be made that "Iron-Mad Wilkinson," as he was widely known, had predecessors who can lay near-equal claim to parenthood. One of them was a luckless clockmaker from Yorkshire named John Harrison, who worked just a few decades earlier to create devices that kept near-perfect time; the other, rather unexpectedly to those who suppose precision to be more or less a modern creation, was a nameless craftsman who worked in Ancient Greece some two thousand years before Harrison, and whose triumph of precise craftsmanship was discovered deep in the Mediterranean at the turn of the last century by a group of fishermen out diving for sponges.

The Greek team, diving in the warm waters south of the Peloponnese, close to the small island of Antikythera, found sponges in abundance, as they usually did. Yet this time they found something else: the spars and tumbled beams of a wrecked ship, most probably a Roman-era cargo vessel. Among all the broken wood, they came upon a diver's dream: a massive trove of marvels of art and luxury, along with, more mysteriously, a telephone directory–size lump of corroded and calcified bronze and wood that was initially discounted and almost discarded as being of little archaeological significance.

Except that after sitting for two years in a drawer in Athens, overlooked and yet all the while patiently drying itself out, the sorry-looking lump fell apart. It sundered itself into three pieces, revealing within, and to the astonishment of all, a mess of more than thirty metallic and cleverly meshing gearwheels. One of these wheels had a diameter almost as wide as the object itself; others were no wider than a centimeter. All had hand-cut triangular teeth—the tiniest wheels had as few as 15; the enormous one had a then-inexplicable 223. It looked as though all the wheels had been cut from a single plate of bronze.

Astonishment at this discovery quickly turned to disbelief, to skepticism, to a kind of puzzled fearfulness among scientists who simply could not believe that even the most sophisticated of Hellenistic engineers had ever been capable of making such a thing. So, for almost half a century, this most intimidating machine—if that is what it was—was locked away again, secured and contained like a deadly pathogen. It was given a name, the Antikythera mechanism, for the island, halfway between Crete and the southern tendrils of mainland Greece, off which it was found. It was then quietly and casually all but erased from a Greek archaeological history that was much more comfortable dealing with the more customary fare of vases and jewelry, amphorae and coins, and statues of marble or the most lustrous bronze. A handful of slim books and pamphlets were published, declaring the device to be some kind of astrolabe or planetarium, but otherwise, there was a near-universal lack of interest in the find.

It was not until 1951 that Derek Price, a young British student of the history and social impact of science, won permission to take a closer look at the Antikythera mechanism, and for the next two decades he subjected the shattered relic, with a total of now more than eighty additionally found bits and pieces as well as the three main fragments, to blizzards of X-rays and wafts of gamma radiation, probing secrets that had been hidden for two thousand years. Eventually, Price decided the work was much more complex and important than a mere astrolabe—it was in fact more likely to be the once-beating heart of a mysterious computing device of unimagined mechanical complexity, one that had evidently been made in the second century BC and was clearly a work of staggering genius.

Price's work in the 1950s was limited by his technology's inability properly to peer inside the device. All this changed with the invention twenty years later of magnetic resonance imaging, or MRI, which led in 2006, more than a century after the sponge seekers made their first find, to the publication in *Nature* of a profoundly more detailed and sophisticated analysis.

The world-scattered team of specialist researchers who produced the *Nature* article concluded that what the Greek divers had pulled to the surface were the remains of a miniaturized and neatly boxed mechanical device, an analog computer, essentially, with dials and pointers and rudimentary instructions for how to use it. It was a device that "calculated and displayed celestial information, particularly cycles such as the phases of the moon and a luni-solar calendar." Moreover, minuscule inscribed lettering in Corinthian Greek chased into the machine's brass work—a total of 3,400 letters, all millimeter-size, have been found thus far—suggested that the gearwheels, once fully engaged with one another with the turning of a crank on the side of the box, could also predict the movement of the five other planets then known to the Ancient Greeks.*

---

* Both Classical and the later Hellenistic Greek astronomers knew of five other planets: Mercury, Venus, Mars, Jupiter, and

Enthusiasts, a small but fervent corps of devotees of this extraordinary little instrument, have since built working models of the mechanism, in wood and brass and, in one instance, with its bronze innards expanded and exploded as in a 3-D checkers game, between layers of transparent Perspex. It was the numbers of teeth on the various wheels that offered the first clues as to how they might have been employed by the machine's makers. The fact that there were 223 teeth on the largest of the gearwheels, for example, provided a eureka moment for the investigators, who remembered that Babylonian astronomers, who were the most astonishingly able watchers of the skies, had calculated that lunar eclipses were usually separated by 223 full moons. Use of this particular wheel, then, would have enabled the user to predict the timing of eclipses of the moon (just as other wheels and combinations of wheels would have turned pointers on dials to display phases and planetary perturbations) and the dates, more trivially, of upcoming public sporting events, most notably the ancient Olympic Games.

Saturn. Although the Greek names for them were different from our own—in order, they were Hermes, Aphrodite, Ares, Zeus, and Cronos—the word *planet* is Greek, and means "wanderer," because, to early eyes, the bodies wandered across the sky in a different manner from the stars behind them.

Modern investigators have concluded that the device was very well made, "with some parts constructed to accuracies of a few tenths of a millimeter." By that measure alone, it would seem that the Antikythera mechanism can lay claim to being a most precise instrument—and, crucially for this introduction to the story, maybe the first precision instrument ever made.

Except that there is an inherent flaw in this claim. The device, as model-tested by the legions of fascinated modern analysts, turns out to be woefully, shamefully, uselessly inaccurate. One of the pointers, which supposedly indicates the position of Mars, is on many occasions thirty-eight degrees out of true. Alexander Jones, the New York University antiquities professor who has perhaps written most extensively about the Antikythera mechanism, speaks of its sophistication as being that only "of a young and rapidly developing craft tradition," and of "questionable design choices" by its makers, who, in summary, produced a device that was "a remarkable creation, but not a miracle of perfection."

There is one additional puzzling aspect of the mechanism that still intrigues historians of science to this day: while it was filled to bursting with what is self-evidently a complicated assemblage of clockwork, none of its assemblers apparently ever thought of using it *as a clock*.

Hindsight permits us to be puzzled, of course, and persuades us to want to reach back to the Greeks and shake them a little for ignoring what to us seems obvious. For time was already being measured in Ancient Greece with the help of all manner of other devices, most popularly with sundials, dripping water, hourglasses (as in egg timers), oil lamps with time-graduated fuel holders, and slow-burning candles with time graduations on the stick. And though the Greeks possessed (as we now know from the existence of the mechanism) the wherewithal to harness clockwork gears and make them into timekeepers, they never did so. The penny never dropped. It never dropped for the Greeks or, subsequently, for the Arabs or, even beforehand, for the much more venerable civilizations of the East. It would take many more centuries for mechanical clocks to be invented anywhere, but once they were, they would have precision as their most essential component.

Though the eventual function of the mechanical clock, brought into being by a variety of claimants during the fourteenth century, was to display the hours and minutes of the passing days, it remains one of the eccentricities of the period (from our current viewpoint, that is) that time itself first played in these mechanisms a somewhat subordinate role. In their earliest medieval incarnations, clockwork clocks, through their employment of complex Antikythera-style gear trains and florid and beautifully crafted decorations and dials, displayed astronomical information at least as an equal to the presentation of time. It was almost as though the passage of celestial bodies across the heavens was considered more significant than the restless ticking of the passage of moments, of that one-way arrow of time that Newton so famously called "duration."

There was a reason for this. Nature's offerings of dawn, midday, and dusk already provided the temporal framework—the mundane business of when it was time to rise and work, when came the time to rest, to mop the brow and take a drink, and when the time to take nourishment and prepare for sleep. The more finicky details of time (a man-made matter, after all), of whether it was 6:15 a.m. or ten minutes to midnight, were necessarily of lesser importance. The behavior of the heavenly bodies was ordained by gods, and therefore was a matter of spiritual significance. As such, it was far worthier of human consideration than our numerical constructions of hours and minutes, and was thus more amply deserving of flamboyant mechanical display.

Eventually, though, the reputation and standing of the hours and minutes themselves did manage to rise through the ranks, did come to dominate the usage of the clockwork mechanisms that became known, generically, as timekeepers. The Ancients may have looked upward to the skies to gather what time it was, but once machinery began to perform the same task, a vast range of devices took over the duty, and has done so ever since.

Monasteries were the first to employ timekeepers, the monks having a need to awaken and observe in some detail the canonical hours, from Matins to Compline by way of Terce, None, and Vespers. And as various other professions and callings started to appear in society (shopkeepers, clerks, men of affairs bent on holding meetings, schoolteachers due to instruct to a rigid schedule, workers on shifts), the need for a more measured knowledge of numerical time pressed ever more firmly. Toilers in the fields could always see or hear the hour on the distant church clock, but city dwellers late for a meeting needed to know how many minutes remained until the "appointed hour" (a phrase that gained currency only in the sixteenth century, by which time public mechanical clocks were widely on display).

On land, it was the railways that most prolifically showed—one might say defined—the employment of time. The enormous station clock was more glanced at than any other feature of the building; the image of the conductor consulting his (Elgin, Hamilton, Ball, or Waltham) pocket watch remains iconic. The timetable became a biblically important volume in all libraries and some households; the concept of time zones and their application to cartography all stemmed from railways' imprint of timekeeping on human society.

Yet, before the chronological influence of railways, there was one other profession that above all others truly needed the most precise timekeeping. It was that which had been developing fast since the European discovery of the Americas in the fifteenth century and the subsequent consolidation of trade routes to the Orient: the shipping industry.

Navigation across vast and trackless expanses of ocean was essential to maritime business. Getting lost at sea could be costly at best, fatal at worst. Also, because the exact determination of where a ship might be at any one moment was essential to the navigation of a route, and because one part of that determination depends, crucially, on knowing the exact time aboard the ship and, even more crucially, the exact time at some other stable reference point on the globe, maritime clockmakers were charged with making the most precise of clocks.★

★ Once out of sight of land, ships' crews have no means of knowing accurately their exact position. To determine their *latitude*, the distance north or south of the equator, is easy, requiring only the measurement of the altitude of the sun at midday or (in the Northern Hemisphere) the pole star at night. But determining

And none was more sedulously dedicated to achieving this degree of exactitude than the Yorkshire carpenter and joiner who later became England's, perhaps
the world's, most revered horologist: John Harrison, the
man who most famously gave mariners a sure means of
determining a vessel's longitude. This he did by painstakingly constructing a family of extraordinarily precise
clocks and watches, each accurate to just a few seconds
in years, no matter how sea-punished its travels in the
wheelhouse of a ship. An official Board of Longitude
was set up in London in 1714, and a prize of twenty
thousand pounds offered to anyone who could determine longitude with an accuracy of thirty miles. It was

---

longitude, the distance east or west around the world from the
ship's home port, is much more difficult. Longitude meridians
mark the *time difference* between places, as the planet turns
through 360 degrees every twenty-four hours, so each hour is
equivalent to a drawn meridian of 15 degrees of longitude. But
the time difference, and thus the longitude, can be worked out
only if the time back at the home port is known by the ship at
sea (its own local time being comparatively easy to determine
from the sun and the stars). And for any timekeeper (on board
a ship rolling violently in storms, passing through areas of fierce
heat and deep cold, and with the clock never once being allowed
to stop) to maintain an accurate record of this time was, to early
eighteenth-century navigators, deemed well-nigh impossible.

John Harrison who, eventually and after a lifetime of heroic work on five timekeeper designs, would claim the bulk of the prize.

Harrison's legacy is much treasured. The curator of the Greenwich Royal Observatory, high on its all-seeing hill above the Maritime Museum to the east of London, comes in each dawn to wind the three great clocks that he and his staff are disposed simply to call "the Harrisons." He stands on much ceremony to wind them, well aware of the immense historical importance invested in the three timepieces and their one unwound sibling. Each was a prototype of the modern marine chronometer, which, in allowing ships to fix their positions at sea with accuracy, has since saved countless sailors' lives. (Before the existence of the marine chronometer, before ships' masters had the ability to determine exactly where they were, vessels tended to collide with importunate frequency into islands and headlands that loomed up unexpectedly before their bows. Indeed, it was the catastrophic collision off the Cornish coast of Admiral Sir Cloudesley Shovell's squadron of warships in 1707 [which drowned him and two thousand of his sailors] that compelled the British government to think seriously about the means of figuring out longitude—setting up a Board of Longitude and offering prize money—which led, ultimately, to the making of the small family of clocks that are wound each dawn at Greenwich.)

There were other reasons for the vast importance of the Harrisons. By allowing ships to know their positions and plot their voyages with efficiency, accuracy, and precision, these clocks and their successors enabled the making of untold trading fortunes. And though it may no longer be wholly respectable to say so, the fact that the Harrison clocks were British-invented and their successor clocks firstly British-made allowed Britain in the heyday of her empire to become for more than a century the undisputed ruler of all the world's oceans and seas. Precise-running clockwork made for precise navigation; precise navigation made for maritime knowledge, control, and power.

And so the curator pulls on his white curatorial gloves and, using in each case a unique double set of brass keys, unlocks the tall glass-sided cabinets in which the great timekeepers stand. Each of the three is on near-permanent loan from Britain's Ministry of Defence. The earliest made, finished in 1735 and known these days as H1, the curator can wind with a single strong downward pull on a chain made of brass links. The later pair, the midcentury H2 and H3, require simply a swift turn of a key.

The final device, the magnificent H4 "sea watch" with which Harrison eventually won his prize money, remains unwound and silent. Housed in a five-inch-diameter silver case that makes it look rather like an enlarged and biscuit-thick version of grandfather's pocket watch, it requires lubrication and, if it runs, will become less precise with time as the oil thickens; it will lose *rate*, as horologists say. Moreover, if H4 were kept running, only its second hand would be seen to be moving, and so, as spectacle, it would be somewhat uninteresting—and as a trade-off for the inevitable wear and tear of the movement beneath, the sight of a moving second hand made no sense. So the decision of the observatory principals over the years has been to keep this one masterpiece in its quasi-virgin state, much like the unplayed Stradivarius violin at the Ashmolean Museum in Oxford,* as an untouched testament to its maker's art.

And what sublime pieces of mechanical art John

---

* Oxford lore has it that this violin, known as Le Messie ("the Messiah"), remained unplayed and virginal until a man from an American southern state arrived and insisted that he be allowed to play it, and wept bitterly when refused. The keeper relented and locked the man in the room for fifteen minutes, during which music of an ethereal beauty such as no one in the museum had ever heard before, wafted through the doors, to the delight of all.

Harrison made! By the time he decided to throw his hat in the ring for the longitude prize, he had already constructed a number of fine and highly accurate timekeepers—most of them pendulum clocks for use on land, many of them long-case clocks, each one more refined than the last. Harrison's skills lay in the imaginative improvement of his timekeepers, rather than in the decorative embellishment that many of his eighteenth-century contemporaries were known for.

He was fascinated, for instance, with the problem of friction, and in a radical departure from the norm, he made all his early clocks with wooden gearwheels, which needed none of the lubricant oils of the day, oils that became notoriously more viscous with age and had the trying effect of slowing down most clockwork movements. To solve this problem, he made all his gear trains first of boxwood and then of the dense, nonfloating Caribbean hardwood *Lignum vitae*, combined in both cases with pivots made of brass. He also designed an extraordinary escapement mechanism, the ticking heart of the clock, that had no sliding parts (and hence no friction, either) and that is still known as a grasshopper escapement because one of the components jumps out of engagement with the escape wheel, just as a grasshopper jumps suddenly out of the grass.

A portable precision clock designed for use on a rolling ship cannot easily use a long gravity-driven pendulum, however, and the first three timepieces Harrison designed for the contest were powered by systems of weights that look very different from the heavy plumb bobs that hang in a conventional long-case clock. They are instead brass bar balances that look like a pair of dumbbells, both placed vertically on the outer edges of the mechanism and its wheel trains, and connected at their tops and bottoms by pairs of springs, which provide the mechanism with a form of artificial gravity, as Harrison wrote. These springs allow the two balance beams to swing back and forth, back and forth, nodding toward and away from each other endlessly (provided that the white-gloved curator, successor on land to the ship's master at sea, winds the mechanism daily) as the clock ticks on.

H1, H2, and H3, each clock a subtle improvement upon its predecessor, each the work of years of patient experimentation—H3 took Harrison fully nineteen years to build—employ essentially the same bar balance principle, and when they are working, they are machines of an astonishing, hypnotic beauty and seemingly bewildering complexity. Many of the improvements that this former carpenter and viola player, bell tuner and choirmaster—for eighteenth-century polymaths were polymaths indeed—included in each have gone on to become essential components of modern precision machinery: Harrison created the encaged roller bearing, for example, which became the predecessor to the ball bearing and led to the founding of huge modern corporations such as Timken and SKF. And the bimetallic strip, invented solely by Harrison in an attempt to compensate for changes in temperature in his H3 timekeeper, is still employed in scores of mundane essentials: in thermostats, toasters, electric kettles, and their like.

As it happened, none of these three fantastical contraptions, however beautiful in appearance and revolutionary in design they may have been, turned out to be a success. Each was taken to a ship and used by the crew as timekeeper, and each time, though the timekeeper offered an improvement on surmising the ships' various positions, the accuracy of the vessel's clock-derived longitude was wildly at variance from what the Board of Longitude demanded—and so no prize was awarded. Harrison's genius and determination were recognized, though, and hefty grants continued to come his way in the hope that he would, in time, make a horological breakthrough. And this he did, at last, when between the four years from 1755 until 1759 he made not another clock, but a watch, a watch that has been known, since it was cleaned and restored in the 1930s, simply as H4.*

The watch was a technical triumph in every sense. After thirty-one years of near-obsessive work, Harrison managed to squeeze almost all the improvements he had engineered in his large pendulum clocks into this single five-inch silver case, and add some others, to make certain that his timekeeper was as close to chronological infallibility as was humanly possible.

In place of the oscillating beam balances that made the magic madness of his large clocks so spectacular

---

* The man who restored (and assigned the initials to) the Harrisons, Rupert Gould, was something of a character. A six-foot-four-inch pipe-smoking former Royal Navy officer known as a congenial children's broadcaster, a scholar of esoteric subjects, a sometime Wimbledon centre court tennis umpire, and an expert on the Loch Ness Monster, he was also famous for violent, drunken outbursts, a number of savage mental breakdowns, and curious sexual predilections, all of which culminated in a spectacular 1927 divorce action that held the nation enthralled. He wrote and illustrated a classic work on seagoing clocks in 1923 (still in print) and soon thereafter managed to persuade the Royal Observatory to release the Harrison clocks, which were decaying in a seldom-visited basement. He got H1 to work again after 165 years. The restoration effort consumed 10 years of his life, a life memorialized in a 2000 TV drama, *Longitude*, in which he was played by the actor Jeremy Irons.

to see, he substituted a temperature-controlled spiral mainspring, together with a fast-beating balance wheel that spun back and forth at the hitherto unprecedented rate of some eighteen thousand times an hour. He also had an automatic *remontoir*, so-called, which rewound the mainspring eight times a minute, keeping the tension constant, the beats unvarying. There was a downside, though: this watch needed oiling, and so, in an effort to reduce friction and keep the needed application of oil to a minimum, Harrison introduced, where possible, bearings made of diamond, one of the early instances of a jeweled escapement.

It remains a mystery just how, without the use of precision machine tools—the development of which will be central to the story that follows—Harrison was able to accomplish all this. Certainly, all those who have made copies of H4 and its successor, K1 (which was used on all Captain James Cook's voyages), have had to use machine tools to fashion the more delicate parts of the watches: the notion that such work could possibly be done by the hand of a sixty-six-year-old John Harrison still beggars belief.

Once his task was completed, he handed the finished watch over to the Admiralty for its crucial test. The instrument (in the care of Harrison's son William, who acted as its chaperone) was taken aboard the HMS *Dept-*

ford, a fifty-gun fourth-rate ship of the line, and sent out on a five-thousand-mile voyage from Portsmouth to Jamaica.* Careful observation at the end of the trip showed the watch to have accumulated a timekeeping error of only 5.1 seconds, well within the limits of the longitude prize. Over the entire 147 days of a voyage that involved a complex and unsettling stormy return journey (in which William Harrison had to swaddle the timekeeper in blankets), the watch error was just 1 minute 54.5 seconds, a level of accuracy never imagined for a seaborne timekeeping instrument.

And while it would be agreeable to report that John Harrison then won the prize for his marvelous creation, much has been made of the fact that he did not. The Board of Longitude prevaricated for years, the Astronomer Royal of the day declaring that a much better way of determining longitude, known as the lunar distance method, was being perfected, and that there was therefore no need for sea clocks to be made. Poor John Harrison, therefore, had to visit King George III (a great admirer, as it happens) to ask him to intercede on his behalf.

A series of humiliations followed. H4 was forced to be

---

* With an unscheduled stop in Madeira to replace the crew's tainted beer supply.

tested once again, and recorded an error of 39.2 seconds over a forty-seven-day voyage—once again, well within the limits set by the Board of Longitude. Harrison then had to dismantle the watch in front of a panel of observers and hand his precious instrument to the Royal Observatory for a ten-month running trial to check (once again, but this time on a stable site) its accuracy. It was torturous and vexing for the now-elderly Harrison, who at seventy-nine was becoming increasingly and understandably embittered by the whole procedure.

Finally, and thanks in large part to King George's intervention, Harrison did get almost all his money. The popular memory of him, though, is of a genius hard done by, and his great clocks and the two sea watches, H4 and K1, remain the most potent memorials, three of them beating out the time steadily and ceaselessly as a reminder of how their maker, with his devotion to precision and accuracy in his craft work, helped so profoundly to change the world.

The Antikythera mechanism, then, was a device remarkable and precise in its making and aspect, but its inaccuracy and understandably amateurish construction rendered it unreliable and, in practical terms, well-nigh useless. John Harrison's timekeepers, though, were both precise and accurate, but given that they took years to make and perfect, and were the result of hugely costly craftsmanship, it would be idle to declare them either as candidates or as the fountainhead for true and world-changing precision. Also, though intending no disrespect to an indelible technical achievement, it is worth noting that John Harrison's clockworks enjoyed perhaps only three centuries' worth of practical usefulness. Nowadays, the brassbound chronometer in a ship's chart room, just like the sextant kept in its watertight morocco box, is a thing more decorative than essential. Time signals of impeccable accuracy now come across the radio. The digital readout of longitude and latitude coordinates come to a ship's bridge from a Global Positioning System's (GPS) interrogation of faraway satellites. Clockwork machines, however beautifully their gears may be cut and enclosed in casings, however precious and intricately engraved, are a creation of yesterday's technology, and are retained nowadays by and large for their precautionary value only: if the seagoing vessel loses all power, or if the master is a purist with a disdain for technology, then John Harrison's works have real practical worth. Otherwise, his clocks gather dust and salt, or are kept in glass cases, and his name will begin to slip gradually astern, to vanish inevitably and soon in a sea fret of history, way stations at the beginning of the voyage.

For precision to be a phenomenon that would entirely alter human society, as it undeniably has done and will do for the foreseeable future, it has to be expressed in a form that is duplicable; it has to be possible for the same precise artifact to be made again and again with comparative ease and at a reasonable frequency and cost. Any true and knowledgeable craftsman (just like John Harrison) may be able, if equipped with sufficient skill, ample time, and tools and material of quality, to make one thing of elegance and evident precision. He may even make three or four or five of the same thing. And all will be beautiful, and most will inspire awe.

Large cabinets in museums devoted to the history of science (most notably at Oxford and Cambridge and Yale) are today filled with such objects. There are astrolabes and orreries, armillary spheres and astraria, octants and quadrants, and formidably elaborate sextants, both mural and framed, which are to be seen in particular abundance, most of them utterly exquisite, intricate, and assembled with a jeweler's care.

At the same time, all of each instrument was perforce made *by hand*. Every gear was hand-cut, as was every component part (every mater and rete and tympan and alidade, for example; astrolabes have their own quite large vocabulary), every tangent screw and index mirror (words relating to sextants are similarly various). Also, the assembly of each part to every other and the adjustment of the assembled whole—all had to be accomplished with, quite literally, fingertip care. Such an arrangement produced fine and impressive instruments, without a doubt, but given the manner in which they were made and how they were put together, they could necessarily have been available only in rather limited numbers and to a small corps d'élite of customers. They may have been precise, but their precision was very much for the few. It was only when precision was created for the many that precision as a concept began to have the profound impact on society as a whole that it does today.

And the man who accomplished that single feat, of creating something with great exactitude and making it not by hand but with a machine, and, moreover, with a machine that was specifically created to create it—and I repeat the word *created* quite deliberately, because a machine that makes machines, known today as a "machine tool," was, is, and will long remain an essential part of the precision story—was the eighteenth-century Englishman denounced for his supposed lunacy because of his passion for iron, the then-uniquely suitable metal from which all his remarkable new devices could be made.

In 1776, the forty-eight-year-old John Wilkinson, who would make a singular fortune during his eighty years of life, had his portrait painted by Thomas Gainsborough, so he is far from an uncelebrated figure—but if not uncelebrated, then not exactly celebrated, either. It is notable that his handsome society portrait has for decades hung not in prominence in London or Cumbria, where he was born in 1728, but in a quiet gallery in a museum far away in Berlin, along with four other Gainsboroughs, one of them a study of a bulldog. The distance suggests a certain lack of yearning for him back in his native England. And the New Testament remark about a prophet being without honor in his own country would seem to apply in his case, as Wilkinson is today rather little remembered. He is overshadowed quite comprehensively by his much-better-known colleague and customer, the Scotsman James Watt, whose early steam engines came into being, essentially, by way of John Wilkinson's exceptional technical skills.

History will show that the story of such engines, which were so central to the mechanics of the following century's Industrial Revolution, is inextricably entwined with that of the manufacture of cannons, and not simply because both men used components made from heavy hunks of iron. A further link can be made, between the thus gun-connected Wilkinson and Watt on the one hand and the clockmaker John Harrison on the other, as it will be remembered that Harrison's early sea clock trials were made on Royal Naval warships of the day, warships that carried cannon in large numbers. Those cannons were made by English ironmasters, of whom John Wilkinson was among the most prominent and, as it turned out, the most inventive, too. So the story properly begins there, with the making of the kind of large weapons used by Britain's navy during the mid-eighteenth century, a time when the nation's sailors and soldiers were being kept exceptionally busy.*

John Wilkinson was born into the iron trade. His father, Isaac, originally a Lakeland shepherd, discovered by fortuitous chance the presence of both ore and coal on his pastures and so became in time an ironmaster,

---

* During Wilkinson's lifetime, the newly created Great Britain was very much in a fighting mood, indulging in such conflicts as the War of Jenkins' Ear, with Spain; the War of the Austrian Succession, against France; the Seven Years' War, with France and Spain together; the American Revolutionary War; the Fourth Anglo-Dutch War; and then, once Ireland joined England and Scotland to make the United Kingdom, the Napoleonic Wars. Wilkinson cannons were used in almost all the major battles.

John "Iron-Mad" Wilkinson, whose patent for boring cannon barrels for James Watt marked both the beginning of the concept of precision and the birth of the Industrial Revolution.

a trade very much of its time. The word describes the owner of a family of furnaces, and one who used them to smelt and forge iron from its ore with either charcoal (which stripped England of too-large tracts of forest) or (as an environmentally responsible response) coal that had been half burned and transmuted into coke. John himself, uncomfortably born, it was said, bumping along in a market cart while his mother was en route to a country fair, became fascinated by white heat and molten metal and the whole process of taking mere rocks that lay underground and creating useful things simply by violently heating and hammering them. He learned the trade at the various places in the English Midlands and the Welsh Marches where his father settled down, and was sufficiently adept that by the early 1760s, by now married into money and owning a considerable foundry in the Welsh-English borderland village of Bersham, he began in earnest the production, according to the firm's first ledger, of "calendar-rolls, malt mill rolls, sugar rolls, pipes, shells, grenades and guns." It was the final item on the list that would give the tiny village of Bersham, along with the man who would become its most prosperous resident and its largest employer, a unique place in world history.

Bersham, which lies in the valley of the River Clywedog, enjoys an indisputable though half-forgotten role both in the founding of the Industrial Revolution and in the story of precision. For it is here that on January 27, 1774, John Wilkinson, whose local furnaces, all fired by coal, were producing a healthy twenty tons of good-quality iron a week, invented a technique for the manufacture of guns. The technique had an immediate cascade effect very much more profound than those he ever imagined, and of greater long-term importance, I would argue, than the much more famed legacies of his friend and rival Abraham Darby III, who threw up the still-standing great Iron Bridge of Coalbrookdale that attracts tourist millions still today, and is regarded by most modern Britons as the Industrial Revolution's most potent and recognizable symbol.

Wilkinson filed a patent, Number 1063—it was still quite early in the history of British patents, which were first issued in 1617—with the title "A New Method of Casting and Boring Iron Guns or Cannon." By today's standards, his "new method" seems almost pedestrian and an all-too-obvious improvement in cannon making. In 1774, however, a time when naval gunnery all over Europe was enjoying a period of sudden scientific improvement in both technique and equipment, Wilkinson's ideas came as a godsend.

Up until then, naval cannons (most particularly the thirty-two-pound long gun, a standard on first-rate ships of the line in the Royal Navy, often ordered a hundred at a time when a new vessel was launched) were cast hollow, with the interior tube through which the powder and projectile were pushed and fired preformed as the iron was cooling in its mold. The cannon was then mounted on a block and a sharp cutting tool advanced into it at the end of a long rod, with the idea of smoothing out any imperfections on the tube's inner surface.

The problem with this technique was that the cutting tool would naturally follow the passage of the tube, which may well not have been cast perfectly straight in the first place. This would then cause the finished and polished tube to have eccentricities, and for the inner wall of the cannon to have thin spots where the tool wandered off track. And thin spots were dangerous—they meant explosions and bursting tubes and destroyed cannon and injuries to the sailors who manned the notoriously dangerous gun decks. The poor quality of early eighteenth-century naval artillery pieces led to failure rates that decidedly alarmed the sea lords at Admiralty headquarters in London.

Then came John Wilkinson and his new idea. He decided that he would cast the iron cannon not hollow but solid. This, for a start, had the effect of guaranteeing the integrity of the iron itself—there were fewer parts that cooled early, for example, as would happen if there was a form installed to create the inner tube. A solid cylindrical chunk of iron, heavy though it might have been, could, if carefully made, come out of the Bersham furnaces without the air bubbles and spongy sections ("honeycomb problems," as they were called) for which hollow-cast cannon were then notorious.

Yet the true secret was in the boring of the cannon hole. Both ends of the operation, the part that did the boring and the part to be bored, had to be held in place, rigid and immovable. That was a canonical truth, as true today as it was in the eighteenth century, for to cut or polish something into dimensions that are fully precise, both tool and workpiece have to be clasped and clamped as tightly as possible to secure immobility. Moreover, in the specific case of gun barrels, there could be no allowable temptation for the boring tool to wander while the bore was being made. This was the reason the cannons were cast solid rather than hollow. To do otherwise was to risk explosive catastrophe.

In the first iteration of Wilkinson's patented process, this solid cannon cylinder was set to rotating (a chain was wrapped around it and connected to a waterwheel) and a razor-sharp iron-boring tool, fixed onto the tip of a rigid base, was advanced directly into the face of the rotating cylindrical workpiece. This created a brand-new hole, straight and precise, as the boring tool was pushed directly into the iron. "With a rigid boring bar and the bearing true," wrote a recent biographer of Wilkinson's, somewhat poetically, "accuracy was bound to ensue." In later versions, it was the cannon that remained fixed and the tool, itself now connected to the waterwheel, that was turned. In theory, and provided that the turning bar itself was rigid; that it was supported at both ends and so maintained its rigidity; and that, as it was advanced into the hole it was boring into, the cylinder face did not bend or turn or hesitate or waver in any way, a hole of great accuracy could be created.

Indeed, that is just what was obtained. Cannon after cannon tumbled from the mill, each accurate to the measurements the navy demanded, each one, once unbolted from the mill, identical to its predecessor, each one certain to be the same as the successor that would next be bolted onto it. The new system worked impeccably from the very start, encouraging Wilkinson to apply for and indeed receive his famous patent.

Instead of an eccentrically drilled-out version of a previously cast hole in a cannon barrel that was already peppered with flaws and weak spots, and which, if it fired at all, would hurl the ball or chain shot or shell wildly through the air, the Royal Navy now received from the Bersham works wagonloads of guns that had a much longer shelf life and would fire their grapeshot or canister shot or explosive shells exactly at their intended target. The improvements were all thanks to the efforts of John Wilkinson, ironmaster. Already a wealthy man, Wilkinson prospered mightily as a result: his reputation soared, and new orders flooded in. Soon, his ironworks alone were producing fully one-eighth of all the iron made in the country, and Bersham was firmly set to be a village for the ages.

Yet what elevates Wilkinson's new method to the status of a world-changing invention, and Bersham's consequent elevation from the local to the world stage, would come the following year, 1775, when he started to do serious business with James Watt. He would then marry his new cannon-making technique, though this time without a brand-new patent, incautiously, with the invention that Watt was just then in the throes of completing, the invention that would ensure that the Industrial Revolution and much else besides and beyond were powered by the cleverly harnessed power of steam.

The principle of a steam engine is familiar, and is based on the simple physical fact that when liquid water is heated to its boiling point it becomes a gas. Because the gas occupies some 1,700 times greater volume than the original water, it can be made to perform work. Many early experimenters realized this. A Cornish ironmonger named Thomas Newcomen was the first to turn the principle into a product: he connected a boiler, via a tube with a valve, to a cylinder with a piston, and the piston to a beam on a rocker. Each time steam from the boiler entered the cylinder, the piston was pushed upward, the beam tilted, and a small amount of work (a very small amount) could be performed by whatever was on the far end of the beam.

Newcomen then realized he could increase the work by injecting cold water into the steam-filled cylinder, condensing the steam and bringing it back to 1/1,700 of its volume—creating, in essence, a vacuum, which enabled the pressure of the atmosphere to force the piston back down again. This downstroke could then lift the far end of the rocker beam and, in doing so, perform real work. The beam could lift floodwater, say, out of a waterlogged tin mine.

Thus was born a very rudimentary kind of steam engine, almost useless for any application beyond pumping water, but given that early eighteenth-century England was awash with shallow mines that were themselves awash with water, the mechanism proved popular and useful to the colliery community. The Newcomen engine and its like remained in production for more than seventy years, its popularity beginning to lessen only in the mid-1760s, when James Watt, who was then employed making and repairing scientific instruments six hundred miles away at the University of Glasgow, studied a model of its workings closely and decided, in a series of moments of the purest genius, that it could be markedly improved. It could be made efficient, he thought. It could possibly be made extremely powerful.

And it was John Wilkinson who helped to make it so—once, that is, Watt had had his strokes of the purest genius. These can be summed up simply enough. For weeks, alone in his rooms in Glasgow, Watt puzzled over a model of the Newcomen engine, a machine famed for being so woefully inadequate, so inefficient, so wasteful of all the heat and energy expended upon it. Watt, patiently trying out various ways to improve on Newcomen's invention, is reported to have remarked wearily that "Nature has a weak side, if only we can find it out."

He finally did so, according to legend, one Sunday in 1765, as he was taking a restorative walk through a park in central Glasgow. He realized that the central inefficiency of the engine he was examining was that the cooling water injected into the cylinder to condense the steam and produce the vacuum also managed to *cool the cylinder itself.* To keep the engine running efficiently, though, the cylinder needed to be kept as hot as possible at all times, so the cooling water should perhaps condense the steam *not in the cylinder but in a separate vessel,* keeping the vacuum in the main cylinder, which would thus retain the cylinder's heat and allow it to take on steam once more. Moreover, to make matters even more efficient, the fresh steam could be introduced at the top of the piston rather than the bottom, with stuffing of some sort placed and packed into the cylinder around the piston rod to prevent any steam from leaking out in the process.

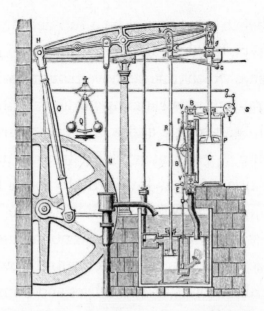

A cross section of a late eighteenth-century Boulton and Watt steam engine. The main cylinder, C, would have been bored by John Wilkinson, the piston, P, fitting snugly inside it to the thickness of an English shilling, a tenth of an inch.

These two improvements (the inclusion of a separate steam condenser and the changing of the inlet pipes to allow for the injection of new steam into the upper rather than the lower part of the main cylinder)—improvements so simple that, at this remove, they seem obvious, even though, to James Watt in 1765, they were anything but— changed Newcomen's so-called fire-engine into a proper and fully functioning steam-powered machine. It became in an instant a device that in theory could produce almost limitless amounts of power.

As he began what would be a full decade of testing and prototype building and demonstrating and seeking funds (during which time he moved south from Scotland to the vibrantly industrializing purlieus of the English Midlands), Watt sought and was swiftly awarded a patent: Number 913 of January 1769. It had a deceptively innocuous title: "A New Invented Method of Lessening the Consumption of Steam and Fuel in Fire-Engines." The modest wording belies the invention's importance: once perfected, it was to be the central power source for almost all factories and foundries and transportation systems in Britain and around the world for the next century and more.

What is especially and additionally noteworthy, though, is that a historic convergence was in the making. For, living and working nearby in the Midlands, and soon to produce a patent himself (the already noted Number 1063 of January 1774, an exact one hundred fifty patents and exactly five years later than James Watt's), was no less an inventor than John Wilkinson, ironmaster.

By then, Wilkinson's amiable madness was making itself felt throughout the ferrous community: all came to learn that he had made an iron pulpit from which he lectured, an iron boat he floated on various rivers, an iron desk, and an iron coffin in which he would occasionally lie and make his frightening mischief. (Women were in plentiful attendance, despite his being a somewhat unattractive man with a massively pockmarked face. He had a vigorous sex drive, fathering a child at seventy-eight by way of a maidservant, a calling of which he was inordinately fond. He kept a seraglio of three such women at one time, each one unaware of the others.)

Still, Wilkinson could and would free himself from these distractions, and by 1775, he and Watt, though of very different temperaments, had met and befriended each other, though it was a friendship based more on commerce than affection. Before long, their two inventions were, and to their mutual commercial benefit, commingled. Wilkinson's "New Method of Casting and Boring Iron Guns or Cannon" was married to Watt's "New Invented Method of Lessening the Consumption of Steam and Fuel in Fire-Engines." It was a marriage, it turned out, of both convenience and necessity.

James Watt, a Scotsman renowned for being pessimistic in outlook, pedantic in manner, scrupulous in affect, and Calvinist in calling, was obsessed with getting his machinery as *right* as it could possibly be. While he was making and repairing and improving the scientific instruments in his workshop in Glasgow, he became well-nigh immured by his passion for exactitude, to much the same degree as had John Harrison in his clock-making workshop in Lincolnshire.

Watt was quite familiar with the early dividing engines and screw thread cutters and lathes and other instruments that were then helping engineers take their first tentative steps toward machine perfection. He was accustomed to instruments that were carefully built and properly maintained, and that worked as they were intended to. He was mortally offended, then, when things went wrong, when inefficiencies were compounded, and when the monster iron engines he was now trying to build in the giant Boulton and Watt factory in Soho performed less well than the brass-and-glass models on which he had experimented back up in Scotland.

His first prototype large engines were spectacular behemoths: thirty feet tall, with a main steam cylinder four feet in diameter and six feet long, a coal-fired boiler, and a separate steam condenser, all massive. All the working parts were connected by a convoluted spiderweb of brass pipes and well-oiled valves and levers, with a spinning two-ball governor that prevented runaways. Above it all was a heavy wooden beam that rocked back and forth with metronomic regularity, turning a huge iron flywheel that in turn worked a pump that gushed water or compressed air or performed other tasks fifteen times a minute. Once at full power, the engine produced a concatenation of noise and heat and a juddering, thudding, stomach-churning intensity that somehow seemed an impossible consequence of merely heating water up to its natural boiling point.

Yet everywhere, perpetually enveloping his engine in a damp, hot, opaque gray fog, were billowing clouds of steam. It was this, this scorching miasma of invisibility, that incensed the scrupulous and pedantic James Watt. Try as he might, do as he could, steam always seemed to be leaking, and doing so not stealthily but in prodigious gushes, and most impudently of all, it was doing so from the engine's enormous main cylinder.

He tried blocking the leak with all kinds of devices, things, and substances. The gap between the piston's outer surface and the cylinder's inner wall should, in theory, have been minimal, and more or less the same wherever it was measured. But because the cylinders were made of iron sheets hammered and forged into a circle, and their edges then sealed together, the gap actually varied enormously from place to place. In some places, piston and cylinder touched, causing friction and wear. In other places, as much as half an inch separated them, and each injection of steam was followed by an immediate eruption from the gap. This is where the blocking came in: Watt tried tucking in pieces of linseed oil–soaked leather; stuffing the gap with a paste made from soaked paper and flour; hammering in corkboard shims, pieces of rubber, even dollops of half-dried horse dung. A solution of sorts came when he decided to wrap the piston with a rope and tighten what he called a "junk ring" around the compressible rope.

Then, by the purest accident, John Wilkinson, in Bersham, asked for an engine to be built for him, to act as a bellows for one of his iron forges—and in an instant, he saw and recognized Watt's steam-leaking problem, and in an equal instant, he knew he had the solution: he would apply his cannon-boring technique to the making of cylinders for steam engines.

So, without taking the precautionary step of filing a new patent for this entirely new application of his method, he proceeded to do with the Watt cylinders exactly what he had done with the naval guns. He had Watt's workmen haul a solid iron cylinder blank the seventy miles across to Bersham. He then strapped the blank (in this case, for the very engine that he, as customer, eventually wanted, so six feet long and thirty-eight inches in diameter) onto a firmly fixed stage, and then secured it with heavy chains to make certain it did not move by so much as a fraction of an inch. He then fashioned a massive cutting tool of ultrahard iron that was three feet across (which should in theory have produced a cut that left a thirty-eight-inch-diameter cylinder with one-inch-thick walls) and bolted it securely to the end of a stiff iron rod eight feet long. This he supported at both ends and mounted onto a heavy iron sleigh that could be ratcheted slowly and steadily into the huge iron workpiece.

As soon as he was ready to begin working the piece, he directed, through a hose, a water-and-vegetable-oil mixture both to cool the thrashing metals and to wash away any fragments of cut iron; opened the water valve for the millrace and wheel that would set the rod and its cutting tool turning; and slowly and steadily, notch by notch by notch, set the rod moving forward until its cutting edge began chewing away at the face of the iron billet.

After just half an hour of searing heat and grinding din, the cylinder was cut. The tool, hot but barely blunted, was withdrawn. The hole, three feet in diameter, looked smooth and clean, straight and true. Using a set of chains and blocks, he placed the heavy cylinder (now rather less heavy, as so much iron had been bored away) upward, on its end. The piston, fractionally less than three feet in diameter itself and smeared with lubricating grease, was carefully lifted up and over the lip of the cylinder and down into its depths.

There was, I like to think, a round of cheers, for the piston slipped noiselessly and snugly into the cylinder and could be lifted up and down with ease and without any apparent leakage of air, of grease, of anything. It then took Watt just a few days, once the disassembled pieces were brought back to his works in Soho, to mount the cylinder in pride of place in what would now be his, and the world's, first working full-scale single-action engine. He and his engineers then added all the supplementary parts (the pipes, the second condenser, the boiler, the rocking arm, the governor, the water tank, the flywheel) and then loaded the firebox with coal, added a primer, lit the fire, and, once the water was hot enough to set steam pouring from the safety line, opened the main valve.

With an enormous chuff-chuff-chuff, the piston began to move up and down, up and down, out of the newly machined cylinder. The rocking beam above then began to oscillate up and back; the connecting rod on the far side started to move up and down, up and down; the set of eccentric sun-and-moon gears on the flywheel started to move; and then the huge wheel itself, tons of solid iron that would in effect store the engine's power, started to turn.

Within moments, with the governor's shiny couplet of balls spinning merrily to keep matters in check, the engine was roaring along at full power, thumping and thudding and whirring and chuffing—and now all perfectly visibly because, for the first time since Watt had begun his experiments, *there was no leaking steam.* The engine was working at maximum efficiency: it was fast, it was powerful, and it was doing just what was demanded of it. Watt beamed with delight. Wilkinson had solved his problem, and the Industrial Revolution—we can say now what those two never imagined—could now formally begin.

And so came the number, the crucial number, the figure that is central to this story, that which appears at the head of this chapter and which will be refined in its exactitude in all the remaining parts of this story. This is the figure of 0.1—one-tenth of an inch. For, as James Watt later put it, "Mr. Wilkinson has bored us several cylinders almost without error, that of 50 inch diameter . . . does not err the thickness of an old shilling at any part." An old English shilling had a thickness of a tenth of an inch. This was the tolerance to which John Wilkinson had ground out his first cylinder.

He might in fact have done even better than that. In another letter, written rather later—by which time Wilkinson had bored no fewer than five hundred cylinders for Watt's engines, which were being snapped up by factories and mills and mines all over the country and beyond—the Scotsman boasted that Wilkinson had "improved the art of boring cylinders so that I promise upon a seventy two inch cylinder being not farther distant from absolute truth than the thickness of an old sixpence at the worst part." An old English sixpence was even slighter: half of a tenth of an inch, or 0.05 inches.

Yet this is a quibble. Whether the thickness of a shilling coin or the thinness of an old sixpence, it does not really matter. The fact is that a whole new world was being created. Machines had now been made that would make other machines, and make them with accuracy, with precision. All of a sudden, there was an interest in tolerance, in the clearance by which one part was made to fit with or into another. This was something quite new, and it begins, essentially, with the delivery of that first machine on May 4, 1776. The central functioning part of the steam engine was possessed of a mechanical tolerance never before either imagined or achieved, a tolerance of 0.1 inches, and maybe even better.

On the far side of the Atlantic Ocean, and precisely two months after the culmination of these events, on July 4, 1776, a whole new political world was to be created. The United States of America was born, with implications unimagined by all.

It was very shortly thereafter that the new nation's principal representative in Europe, Thomas Jefferson, heard tell of these miraculous mechanical advances and started to ponder how his own faraway country might well take advantage of developments that appeared to him to have the very greatest potential.

Maybe, Jefferson declared, they could form the basis for a new trade well suited to his new country. Maybe, replied the engineers in response, we can do better than we have done already, and using their own arcane language of numbers, they translated their ambitions: maybe we can make and machine and manufacture metal pieces in America to a tolerance much greater than John Wilkinson's 0.1. Maybe we can be adroit enough to reach down to 0.01. Maybe better than that—maybe to 0.001. Who could possibly know? As with the new nation, these visionary engineers wondered, so perhaps with the new machines.

As it happened, the engineers—in England, mainly, but also, and most significantly for the next part of the story, in France—would do a great deal better than they ever supposed. The genie of accuracy was now out of the bottle. True precision was now out of the gate, and moving fast.

Chapter 2

(Tolerance: 0.0001)

Extremely Flat and Incredibly Close

It is to the exactitude and accuracy of our machine tools that our machinery of the present time owes its smoothness of motion and certainty of action.

—SIR WILLIAM FAIRBAIRN, BT. (1862), *Report of the British Association for the Advancement of Science*

On the north side of London's Piccadilly, over-looking Green Park and sandwiched between the quarters of the aged and imperturbable Cavalry Club to the west and a rather more ephemeral Peruvian-style ceviche restaurant on its eastern side, stands Number 124, these days an elegant but somewhat anonymous structure providing offices for the discreet and service apartments for the wealthy.

Since 1784, when this far-western end of the great boulevard was still ripe for development, the address had been the home and atelier of a cabinet, engine, and lock maker named Joseph Bramah. On fair-weather days some six years after its opening, when Bramah and Company was an established and familiar small firm, modest gatherings of curious passersby would assemble outside to peer into the front bow window, puzzling at a challenge so difficult that it went unanswered for more than the sixty subsequent years.

There was just a single object on view in the window, placed on a velvet cushion like a religious icon. It was a padlock, oval shaped, of modest size, and with a smooth and uncomplicated external appearance. On its face was written, in a small script legible only to those who pressed their faces close to the window glass, the following words: THE ARTIST WHO CAN MAKE AN INSTRUMENT THAT WILL PICK OR OPEN THIS LOCK SHALL RECEIVE 200 GUINEAS THE MOMENT IT IS PRODUCED.

The designer of this boastfully unbreakable lock was the firm's principal, Joseph Bramah. Its maker, however, was not Bramah but a then-nineteen-year-old former blacksmith's apprentice named Henry Maudslay, whom Bramah had taken on the previous year, entirely because

Joseph Bramah, locksmith extraordinaire, also invented the fountain pen, a device for keeping beer cool and under pressure in a pub basement, and a machine for counting banknotes.

of Maudslay's reputation for having a formidable skill in delicate machining.

It would not be until 1851 that the Bramah lock was successfully—although, as we shall see in a later chapter, controversially—picked and the very handsome pledge★ redeemed. And in the years leading up to this event (which only their descendants would survive to witness), these two men, Bramah and Maudslay, proved themselves to be engineers supreme. They invented all manner of intriguing new devices, and they effectively and independently wrote the rule books for the precise world that was beginning to emerge as a consequence of (or, at least, in the wake of) John Wilkinson's achieve-

---

★ Worth about the price of a small Mercedes today.

ments with his cylinder-boring machine at Bersham. Some of the two men's inventions have faded away into history; some others, however, have survived as the foundations on which much of today's most sophisticated engineering achievements would eventually be built.

Though Maudslay remains today the better-known figure, with a legacy recognized by most engineers, Bramah was at the time perhaps the more showily ingenious of the pair. His first invention was dreamed up while he lay in bed after a fall, and must rank as the least romantic: for a London population that sorely needed an improvement in public hygiene, he built water closets, and he patented his ideas for a system of flaps and a float and valves and pipes that made the device both self-cleansing (flushing, indeed, for the first time) and free from the usual risk of freezing in winter that created unpleasant results for all. He made a small fortune from this creation, selling six thousand in the first twenty years of production, and a Bramah WC was still the centerpiece of the civilized English middle-class bathroom right up until Victoria's Jubilee, a hundred years later.

Bramah's interest in locks, which required far more intricacy and precise workmanship than a toilet, of course, seems to have started when he was elected in 1783 a member of the newly formed (and still there, in its original home) Royal Society for the Encouragement of Arts, Manufactures and Commerce.* What is now simply the Royal Society of Arts, the RSA, back in the eighteenth century had six divisions: Agriculture, Chemistry, Colonies and Trade, Manufactures, Mechanicks (spelled thus), and most quaintly, the Polite Arts. Bramah not unnaturally opted to attend most of the Mechanicks meetings and, soon after joining, rocketed to prominence by the simple act of picking a lock. Not so simply, actually: in September 1783, a Mr. Marshall had submitted for consideration what he declared was a formidably unpickable lock, and had a local expert named Truelove worry away at it with a quiverful of special tools for an hour and a half, before accepting defeat. Then, from the back of the audience stepped Joseph Bramah, who quickly fashioned a pair of instruments and opened the lock in fifteen minutes flat. A buzz of excitement went around the room: they were clearly in the presence of a most Mechanickal man.

Locks were a British obsession at the time. The social and legislative changes that were sweeping the country

---

* Among those who recognized the young Yorkshireman's talent was a surgeon named John Sheldon, who was an expert in embalming, claimed to have been the first Londoner to fly in a balloon, and traveled to Greenland to experiment on a new technique of catching whales by spearing them with harpoons tipped with the poison curare.

in the late eighteenth century were having the undesir-
able effect of dividing society quite brutally: while the
landed aristocracy had for centuries protected itself in
grand houses behind walls and parks and ha-has, and
with resident staff to keep mischief at bay, the enriched
beneficiaries of the new business climate were much
more accessible to the persistent poor. They and their
possessions were generally both visible and, especially
in the fast-growing cities, nearby; they tended to live
in houses and on streets within earshot and slingshot of
the vast armies of the impoverished. Envy was abroad.
Robbery was frequent. Fear was in the air. Doors and
windows needed to be bolted. Locks had to be made,
and made well. A lock such as Mr. Marshall's, pickable
in fifteen minutes by a skilled man, and by a desper-
ate and hungry man maybe in ten, was clearly not good
enough. Joseph Bramah decided he would design and
make a better one.

He did so in 1784, less than a year after picking the
Marshall lock. His patent made it almost impossible for
a burglar with a wax-covered key blank, the tool most
favored by the criminals who could use it to work out
the position of the various levers and tumblers inside
a lock, to divine what was beyond the keyhole, inside
the workings. Bramah's design, which he patented that
August, had the various levers inside a lock rise or fall

to different positions when the key was inserted and turned to release the bolt, but then had those same levers *return to their initial positions* once the bolt had been shot. The effect of this was to make the device almost burglar-proof, for no amount of foraging with a wax key blank would ever allow a picklock to work out where the levers needed to be (as they weren't there anymore) in order to free the bolt.

Once Bramah had come up with this basic mechanical premise, it remained for him, with great cleverness and elegance, to form the entire lock into a cylindrical shape, with its levers not so much rising and falling under the influence of gravity as moving in and out along the radii of the cylinder under the impress of the key's various teeth, and then moving back to their original positions with the aid of a spring, one for each lever. The entire lock could thus be rendered as a small tube-shaped brass barrel, which could be easily fitted into a tube-shaped cavity in a wooden door or an iron safe, and with the deadbolt flush to the door's outer edge (when the lock was open) or settled into its brass cavity in the door frame (when securely closed).

Joseph Bramah would go on to invent many more contraptions and concepts during his life, many of them having nothing to do with locks, but involving his particular other fascination with the behavior of liquids when subjected to pressure. He invented the hydraulic press, for example, with its vast importance in industry worldwide. More trivially, he launched onto the market a primitive form of fountain pen* and drew designs for a propelling pencil; more lastingly, he made the beer engine, which is still employed by the more tradition-ally minded innkeepers, and which would allow beer kept cool in a cellar to be pressure-delivered to thirsty customers in the bar above. (This invention obviated the need for the bartender to stagger up and down the cellar stairs, lugging fresh barrels of ale.) Draft beer drinkers today have little cause to remember the name "Bramah," though there is a pub in Lancashire named for him. Likewise, few banknote printers know that it was Joseph Bramah who made the first machine that could cleverly ensure that their thousands of identical bills each bore a different sequential number. He also made an engine for planing large wooden planks, an-other for making paper, and he forecast that, one day, large screws would be used to propel big ships through the water.

---

* He rather hedged his bets, though, by also inventing a de-vice that could cut multiple pen nibs from a single goose feather quill. If his newfangled metal-nibbed pen with its squeezable rubber ink reservoir didn't catch on, he could always fall back on a mass-produced version of the traditional writing instrument.

Yet it is really only by way of his lock making that Bramah's name has now formally entered the English language. True, one can still find in literature references to a Bramah pen and a Bramah lock—the Duke of Wellington wrote admiringly of each, as did Walter Scott and Bernard Shaw. Yet when the word is used alone—and Dickens did on numberless occasions, in *The Pickwick Papers*, in *Sketches by Boz*, in *The Uncommercial Traveller*—it is a reminder that at least for the Victorian citizenry, his was an eponym: one used a Bramah to open a Bramah, one's home was secured with a Bramah, one gave a Bramah to a favored friend so he or she might visit at all hours, come what may. Only when Mr. Chubb and Mr. Yale arrived on the scene (noted by the *Oxford English Dictionary* as first making it into the language in 1833 and 1869, respectively) did Joseph Bramah's lexical monopoly hit the buffers.

What made a Bramah lock so good was its vastly complicated internal design, of course, but what made it so lastingly good was the precision of its manufacture. And that was less the work of its inventor than of the man—the boy, really—whom Bramah hired to make copious numbers of his device and to make them well, to make them fast, and to make them economically. Henry Maudslay was eighteen years old when Bramah lured him away as an apprentice: he would go on to become one of the most influential figures in the early days of precision engineering, his influence being felt to this day both in his native Britain and around the world.

The very young Maudslay, "a tall, comely young fellow" by the time Bramah hired him, cut his teeth in the Woolwich Royal Arsenal in East London. Working first as a twelve-year-old powder monkey—small boys, fleet of foot, were used by the Royal Navy to bring gunpowder down from the ships' magazines to the gun deck—he was then moved to the carpenter's shop, only to pronounce himself bored by the inaccuracy of wood. It was starkly clear to all who employed him that the youngster much preferred metal. They looked away when he smuggled himself into the dockyard smithy, and they said nothing when he developed a sideline in making a range of useful and very handsome trivets out of cast-off iron bolts.

In 1789, Joseph Bramah cut an anxious figure. The political situation across the Channel was causing an influx of terrified French refugees, most of them bound for London, where the more nervously xenophobic residents of England's capital suddenly started to demand ever more security for their homes and businesses. Bramah, with his patent-protected monopoly, was caught in a bind: he alone could make his locks, but neither he nor any engineer he could find had the ability to make them in sufficient numbers at a low enough price. Most men who called themselves engineers may have been adept at the cruder crafts—at thumping ingots of heat-softened iron with heavy hammers and then working to shape the crudely formed results with anvils, chisels, and, most especially, files—but few had a great feel for delicacy, for the construction of (and the word had only recently been adopted) *mechanisms*.

Change was coming, though. Workers at the smithies of eighteenth-century London were a close-knit group, and word eventually did reach Bramah that a particular youngster at Woolwich was startlingly unlike his older peers and, rather than bashing hunks of iron, was apparently crafting metal pieces of an unusual, fastidious daintiness. Bramah interviewed the teenage Maudslay. Though taking to him immediately, the former was only too well aware that the custom was for any would-be entrant to the trade to serve a seven-year apprenticeship. However, commercial need trumped custom: with would-be patrons beating down his door back on Piccadilly, Bramah had no time to spare for the niceties, decided to take a chance, and hired the youngster on the spot. His decision was to change history.

Henry Maudslay turned out to be a transformative figure. First of all, he solved Bramah's supply problems in an inkling—but not by the conventional means of hiring workers who would make the locks one by one through the means of their own craftsmanship. Instead, and just like John Wilkinson two hundred miles west and thirteen years earlier, Maudslay created a machine to make them. He made a machine tool: in other words, a machine to make a machine (or, in this case, a mechanism). He built a whole family of machine tools, in fact, that would each make, or help to make, the various parts of the fantastically complicated locks Joseph Bramah had designed. They would make the parts, they would make them fast and well and cheaply, and they would make them without the errors that handcrafting and the use of hand tools inevitably bring in their train. The machines that Maudslay made would, in other words, make the necessary parts with precision.

Three of his lock-making devices can be seen today in the Science Museum in London. One is a saw that cut the slots in the barrels; another—perhaps less a machine tool than a means of ensuring that production went along at high speed, with every part made exactly the same—is a quick-grip, quick-release vise, a *fixture* that would hold the bolt steady while it was milled by a series of cutters mounted on a lathe; and the third is a particularly clever device, powered by a foot-operated treadle, that would wind the lock's internal springs and hold them under tension as they were positioned and secured in place until the outer cover, a well-shined brass plate with the flamboyant signatures of the Bramah Lock Company of 124 Piccadilly, London, inscribed on its face, was bolted on to finish the job.

A fourth and, some would argue, most supremely important machine tool component also started to make its widespread appearance around this time. It would shortly become an integral part of the lathe, a turning device that, much like a potter's wheel, has been a mechanical aid to the betterment of human life since its invention in pharaonic Egypt. Lathes evolved very slowly indeed over the centuries. Perhaps the biggest improvement came in the sixteenth century, with the concept of the *leadscrew.* This was a long and (most often, in early times) wooden screw that was mounted under the main frame of the lathe and could be turned by hand to advance the movable end of the lathe toward or away from the fixed end. It could do so with a degree of precision; one turn of the handle might advance the movable part of the lathe by an inch, say, depending on the pitch of the leadscrew. It gave wood turners working on a lathe a much greater degree of control, and allowed them to produce things (chair legs, chess pieces, handles) of great decorative beauty, symmetric loveliness, and baroque complexity.

H. MAUDSLAY.

Henry Maudslay, once a "tall, comely fellow," machined the
innards of Bramah's locks and went on to become the founding
father of precision toolmaking, mass production, and the key
engineering concept of achieving perfect flatness.

Henry Maudslay then improved the lathe itself by
many orders of magnitude—first by making it of iron,
forging its structure stoutly and heavily, and at a stroke
allowing it not merely to machine wooden items, but
also to create symmetry out of shapeless billets of hard
metal, which the flimsy lathes of old were incapable of
doing. This alone might have been sufficient for us to
remember the man, but then Maudslay employed one
further component on his working lathes, a component
whose origins are debated still, however, with the tenor

of the debate pointing to an endless argument that complicates the historiography of precision and precision engineering.

Specifically, the device in question mounted on Maudslay's lathes is known as a slide rest, a part that is massive, strongly made, and securely held but movable by way of screws, and is intended to hold any and all of the cutting tools. It is filled with gears that allow for the adjustment of the tool or tools to tiny fractions of an inch, to permit the exact machining of the parts to be cut. The slide rest is necessarily placed between the lathe's headstock (which incorporates the motor and the mandrel that spins the workpiece around) and the tailstock (which keeps the other end of the workpiece secure). The leadscrew—Maudslay's was made of metal, not wood, and with threads much closer together and with a more delicate pitch than was possible for a wooden version—advances the workpiece. The tools held on the slide rest can then be moved across the path of travel dictated by the leadscrew, thereby allowing the tools to make holes in the workpiece, or to chamfer it or (in due course, once milling had been invented, a process related in the next chapter) mill it or otherwise shape it to the degree that the lathe operator demands. So the leadscrew moves the workpiece longitudinally, and the slide rest that holds the tools that cut or chamfer or make holes in the workpiece moves transversely, or in all sorts of directions that are across the path made by the leadscrew.

Metal pieces can be machined into a range of shapes and sizes and configurations, and provided that the settings of the leadscrew and the slide rest are the same for every procedure, and the lathe operator can record these positions and make certain they are the same, time after time, then every machined piece will be the same—will look the same, measure the same, weigh the same (if of the same density of metal) as every other. The pieces are all replicable. They are, crucially, interchangeable. If the machined pieces are to be the parts of a further machine—if they are gearwheels, say, or triggers, or handgrips, or barrels—then they will be *interchangeable parts*, the ultimate cornerstone components of modern manufacturing.

Of equally fundamental importance, a lathe so abundantly equipped as Maudslay's was also able to make that most essential component of the industrialized world, the screw.

Over the centuries, there were many incremental advances in screw making, as we shall see, but it was Henry Maudslay (once he had invented or mastered or improved or in some other manner become intimately associated with the slide rest on his lathe) who then devised a means of cutting metal screws, efficiently, precisely, and fast. Much as Bramah had a lock in his workshop window on Piccadilly, for reasons of pride as much as for his famous challenge, so Maudslay, Sons and Field placed in the bow window of the firm's first little workshop, on Margaret Street in Marylebone, a single item of which the principal was most proud—and that was a five-foot-long, exactly made, and perfectly straight industrial screw made of brass.

Technically, Maudslay was not the first to perfect a screw-making lathe. Twenty-five years earlier, in 1775, Jesse Ramsden, a scientific instrument maker in Yorkshire who was funded by the same Board of Longitude for which the clockmaker John Harrison had labored, and who was not allowed to patent his invention, had made a small and exquisite screw-cutting lathe. This could cut tiny screws with as many as one hundred twenty-five turns to the inch—meaning it would take one hundred twenty-five turns to advance the screw by one inch—and so would allow the tiniest adjustments to any device to which the screw was harnessed. But Ramsden's was effectively a one-off machine, as delicate as a watch, meant for work with telescopes and navigational instruments, and in no way destined for the making of large-scale devices made of much metal and that could work at great speed and maintain accuracy and be durable. What Maudslay had done with his fully equipped lathe was to create an engine that, in the words of one historian, would become "the mother tool of the industrial age."

Moreover, with a screw that was made using his slide rest and his technique, and with a lathe constructed of iron and not with the wooden frame he and Bramah had used initially, he could machine things to a standard of tolerance of *one in one ten-thousandth of an inch*. Precision was being born before all London's eyes.

So, whoever did invent the slide rest can take the credit for the later precise manufacture of countless components of every conceivable size and shape and relevance to a million and one machined objects. The slide rest would allow for the making of myriad items, from door hinges to jet engines to cylinder blocks, pistons, and the deadly plutonium cores of atomic bombs—as well as, of course, the screw.

But just who did invent it? Not a few say Henry Maudslay, and that he did so in Joseph Bramah's "secret workplace [which] contained several curious machines . . . constructed by Mr. Maudslay with his own hands." Others say it was Bramah. Still others refute the idea of Maudslay's involvement entirely, saying definitively that he did not invent it, nor ever claimed to have done so. Encyclopedias say the first slide rest was actually German, having been seen illustrated in a manuscript in 1480. Andrey Nartov, the Russian scientist who had the eighteenth-century title of personal craftsman to Tsar Peter the Great, was revered as the greatest teacher of lathe operation in Europe (and taught the methods to the then-king of Prussia) and is said to have made a working slide rest (and taken it to London to show it off) as early as 1718. And just in case the story from St. Petersburg has any doubters, a Frenchman named Jacques de Vaucanson quite provably made one in 1745.

Chris Evans, a professor in North Carolina who has written extensively about the early years of precision engineering, notes the competing claims, and cautions against the "heroic inventor" treatment of the story. Far better to acknowledge, he says, that precision is a child of many parents, that its advances invariably overlap, that there are a great many indeterminate boundaries between the various disciplines to which the word *precision* can be attached, and that it was, in its early days, a phenomenon that evolved steadily over three centuries of ever-lessening bewilderment. It is, in other words, a story far less precise than its subject.

That being said, Henry Maudslay's principal legacy is a wholly memorable one, for other inventions and involvements followed his association with Joseph Bramah, from whose employ he left, in a huff, after his request for a raise—he was making thirty shillings a week in 1797—was turned down too curtly for his taste.

Maudslay promptly proceeded to free himself from the circumscribed world of West London lock making, and he entered—one might say, he inaugurated—the very different world of mass production. He created in the process the wherewithal for making, in truly massive numbers, a vital component for British sailing ships. He built the wondrously complicated machines that would, for the next one hundred fifty years, make ships' pulley blocks, the essential parts of a sailing ship's rigging that helped give the Royal Navy its ability to travel, police, and, for a while, rule the world's oceans.

This all came about in a moment of the happiest chance, and just as with Bramah and the lock in Piccadilly, it involved a shopwindow (Henry Maudslay's) and the proud public showroom display of the five-foot-long brass screw Maudslay had made on his lathe and which he had placed there, center stage, as an advertisement of his skills. Soon after he set up the screw display, so naval legend has it, came the serendipitous moment. It involved the two figures who were going to create the pulley block factory, and who vowed to do so properly, to fill an urgent and growing need.

A block-making factory of sorts had already been set up in the southern dock city of Southampton in the mid-eighteenth century, performing some of the sawing and morticing of the wooden parts, but much of the finishing work still had to be done by hand, and in consequence, the supply chain remained unreliable at best. And a reliable supply chain was seen to be vital for England's survival.

Britain had been at war with France, on and off, for much of the late eighteenth century, and the arrival on the scene of Napoleon Bonaparte in the aftermath of the French Revolution convinced London that her forces needed to be at the ready for much of the early nineteenth century, too. Of the two British fighting forces, the army and the Royal Navy, it was the admirals who took the lion's share of the war budget, and Britain's docks were soon bristling with big ships ready to cast off at a moment's notice to give any French opponents, Napoleon's especially, a taste of the lash. Shipyards were busy building, dry docks were busy repairing, and the seas from the Channel to the Nile, from the Barbary Coast to Coromandel, were alive with great British men-o'-war, powerful and watchful, ceaselessly on the prowl.

These were, of course, all sailing vessels. Mostly they were enormous craft with wooden hulls and copper-sheathed keels, with three decks ranged with cannon, with enormous masts of Norfolk Island pine supporting equally vast acreages of canvas sailcloth. And all the sail ware of the time were bolts of canvas suspended, supported, and controlled by way of endless miles of rigging, of stays and yards and shrouds and footropes, most of which had to pass through systems of tough wooden pulleys that were known simply to navy men as blocks—pulley blocks, part of a warship's arrangements known within and beyond the maritime world as block and tackle.

A large ship might have as many as fourteen hundred pulley blocks, which were of varying types and sizes depending on the task required. A block with a single pulley might be all that was needed to allow a sailor to hoist a topsail, say, or move a single spar from one location to another. The lifting of a very heavy object (an anchor, for example) might need an arrangement of six blocks, each with three sheaves, or pulleys, and with a rope passing through all six such that a single sailor might exert a pull of only a few easy pounds in order to lift an anchor weighing half a ton. Block-and-tackle physics, taught still in some good primary schools, shows how even the most rudimentary pulley system can offer the greatest of mechanical advantage, and combines this power with an equally great degree of simplicity and elegance.

Blocks for use on a ship are traditionally exceptionally strong, having to endure years of pounding water, freezing winds, tropical humidity, searing doldrums heat, salt spray, heavy duties, and careless handling by brutish seamen. Back in sailing ship days, they were made principally of elm, with iron plates bolted onto their sides, iron hooks securely attached to their upper and lower ends, and with their sheaves, or pulleys, sandwiched between their cheeks, and around which ropes would be threaded. The sheaves themselves were often made of *Lignum vitae*, the very same hard and self-lubricating wood that John Harrison used for the gear trains of some of his clocks: most modern blocks have aluminum or steel sheaves and are themselves made of metal, except where the desired look of the boat is old-fashioned, in which case there is much showy brassware and varnished oak.

Hence the early nineteenth-century Royal Navy's acute concern. An increasingly fractious Napoleonic France lay just twenty miles away across the Channel, and countless maritime problems were demanding Britain's maritime attentions elsewhere: what principally concerned the admirals was not so much the building of enough ships but the supply of the vital blocks that would allow the sailing ships, to put it bluntly, to sail. The Admiralty needed one hundred thirty thousand of them every year, of three principal sizes, and for years past, the complexity of their construction meant that they could be fashioned only by hand. Scores of artisanal woodworkers in and around southern England were originally bent to the task, a supply system that proved notoriously unreliable.

As hostilities at sea became ever more commonplace, as more and more ships were ordered, the drumbeat for a more efficient system became ever louder. The then–inspector general of naval works, Sir Samuel Bentham, finally decided he would act; he would sort things out. And in 1801, Bentham was approached by a figure named Sir Marc Brunel, who said he had in mind a specific scheme for doing so.*

Brunel, a royalist refugee from the very French instability currently so vexing the Lords of the Admiralty— though he had first immigrated to America and

---

* Both Bentham and Brunel had close relatives much more famous than they. Samuel's older brother was Jeremy Bentham, the distinguished philosopher, jurist, and prison reformer whose fully clothed remains, his *auto-icon*, are still seated in a chair in University College London. Brunel's son was the memorably named Isambard Kingdom Brunel, builder of so much that remains spectacularly Victorian in today's Britain, and a popular hero still, ranked by the adoring British public along with Nelson, Churchill, and Newton.

become New York's chief engineer before returning to England to marry—had sized up the mechanics of the block-making problem. He knew the various operations that were necessary to make a finished block—there were at least sixteen of them; a block, simple though it might have looked, was in fact as complex to make as it was essential to employ—and he had roughed out designs for machines that he thought could perform them.★ He sought and, in 1801, won a patent: "A New and Useful Machine for Cutting One or More Mortices

---

★ A block has four basic parts: the wooden shell, the hardwood sheave, a pin for holding the sheave in the shell, and a bushing (the "coak" mentioned in the patent) to minimize wear on the pin. All four would be run hard, of course, every time a rope was passed through the sheave for one of the many reasons a sailor might employ a block. Making the shell alone required seven separate procedures: wooden slices had to be cut from an elm log; the slices had to be cut into rectangles; a hole had to be bored for the pin; mortices had to be cut to allow the sheaves to be inserted; the block's corners had to be cut off and the edges chamfered; the block faces had to be curved and shaped and smoothed; and finally, grooves had to be scored into the block faces to allow ropes to be sited to hold each block secure. Then again, six very different actions needed to be performed on the wooden sheaves, four more were needed for the pins, and two further still for making the bushing. And the entire confection had to be assembled, smoothed off, and sent to storage.

Forming the Sides of and Cutting the Pin-Hole of the Shells of Blocks, and for Turning and Boring the Shivers, and Fitting and Fixing the Coak Therein."

His design was, in more ways than one, revolutionary. He had one machine perform two separate functions—a circular saw, for example, could perform the duties of a mortice cutter as well. He had the surplus motion of one machine drive its neighbor, maintaining a kind of mechanical lockstep. The necessary coordination of the machines one with the other required that the work each machine performed be accomplished with the greatest precision, for a wrong dimension passed into the system by one wrongly set machine would act much as a computer virus does today, amplifying and worsening by the minute, ultimately infecting the entire system, and forcing it to shut own. And rebooting a system of enormous iron-made steam-powered machines with flailing arms and whirling straps and thundering flywheels is not just a matter of pressing a button and waiting half a minute.

Given the complexity of the system he had sold to the navy, it was essential only for Brunel to find an engineer who would and could construct such a set of never-before-made machines, and ensure that they were capable of the repetitious making, with great precision, of the scores of thousands of the wooden pulley blocks the navy so keenly needed.

This is where Henry Maudslay's window comes in. An old friend of Brunel's from his French days, another migrant, named M. de Bacquancourt, happened to pass by the Maudslay workshop on Margaret Street and saw, prominent in the bow window, the famed five-foot-long brass screw that Maudslay himself had made on his lathe. The Frenchman went inside, spoke to some of the eighty employees in the machine shop, and then to the principal himself, and came away firm in the belief that if one man in England could do the work Brunel needed, here he was.

So Bacquancourt told Brunel, and Brunel met Maudslay out at Woolwich. As part of the interview, Brunel then showed the youngster an engineering drawing of one of his proposed machines—whereupon Maudslay, who was able to read drawings in the same way that musicians can read sheet music with the facility that others read books, recognized it in an instant as a means of making blocks. Models of the proposed engines were constructed to show the Admiralty just what was envisioned, and Maudslay set to work, with a formal government commission.

He was to devise and build, as specified in Brunel's drawings, the first precision-made machines in the world that would be established for the sole purpose of manufacturing items. In this case it was pulley blocks, but the items could just as well have been guns, or clocks, or, in time to come, cotton gins or motorcars—en masse.

The project took him six years. The navy built an enormous brick structure in its dockyard at Portsmouth to accommodate the armada of engines they knew was coming. And one by one, first from his workshop back up on London's Margaret Street and then, as the company expanded, from a site in Lambeth, south of the River Thames, Maudslay's epoch-making machines started to arrive.

There would be forty-three of them in total, each performing one or another of the sixteen separate tasks that transformed a felled elm tree into a pulley block to be sent to the naval warehouse. Each machine was built of iron, to keep it solid and sturdy and able to perform its allotted task with the kind of accuracy the navy contract demanded. So there were machines that sawed wood, that clamped wood, that morticed wood, that drilled holes and tinned pins of iron and polished surfaces and grooved and trimmed and scored and otherwise shaped and smoothed the blocks' way to completion. A whole new vocabulary was suddenly born: there were ratchets and cams, shafts and shapers, bevels and worm gears, formers and crown wheels, co-axial drills and burnishing engines.

And all inside the Block Mills, as the structure

was named in 1808, which was soon set to thundering activity. Each of Maudslay's machines was sent power by ever-rotating and flapping leather belts, which themselves were spinning by their connection to long iron axles mounted to the ceiling and that, in turn, were set eventually rotating by an enormous thirty-two-horsepower Boulton and Watt steam engine that roared and steamed and smoked outside the building, in its own noisy and dangerous three-story lair.

The Block Mills still stand as testament to many things, most famously to the sheer perfection of each and every one of the hand-built iron machines housed inside. So well were they made—they were masterpieces, most modern engineers agree—that most were still working a century and a half later; the Royal Navy made its last pulley blocks in 1965. And the fact that many of the parts—the iron pins, for example—were all made by Maudslay and his workers to exactly the same dimensions meant that they were interchangeable, which had implications for the future of manufacturing more generally—as we shall soon see, when the concept of interchangeability was recognized by a future American president.

But the Block Mills are famous for another reason, one with profound social consequences. It was the first factory in the world to have been run entirely from the output of a steam engine. True, earlier machines had been driven by water, and so the concept of mechanization itself was not entirely new. But the scale and the might of what had been built in Portsmouth were different, and stemmed from a source of power not dependent on season or weather or on any external whim. Providing there was coal and water, and an engine made to specifications demanding of the greatest precision, the factory powered by it would run.

The saws and the morticing devices and the drills of the future would thus be powered by engines. These engines would (both here in Portsmouth and then very soon thereafter in a thousand other factories elsewhere, making other things by other means) no longer be turned and powered and manipulated by men. The workers who in their various wood shops had hitherto cut and assembled and finished the navy's pulley blocks had now become the first victims of machinery's cool indifference. Where more than a hundred skilled craftsmen had once worked, and had filled, just, the navy's insatiable appetite, now this thundering factory could feed it with ease, without ever breaking a sweat: the Portsmouth Block Mills would turn out the required one hundred thirty thousand blocks each year, one finished block every minute of every working day, and yet it required a crew of just ten men to operate it.

Precision had created its first casualties. For these were men who needed no special skills. They did no more than feed logs into the slicing machine hoppers and, eventually, take the finished blocks away and stack them inside the storehouses; or else they took their oilcans and their bunches of cotton waste and set to greasing and lubricating and polishing and keeping a weather eye on the clanging and clattering maelstrom of black-and-green and brass-trimmed behemoths, all endlessly mocking them, by revolving and spinning and belching and rocking and lifting and splitting and sawing and drilling, an immense orchestra of machinery that was crammed into the massive new building.

The social consequences were immediate. On the plus side of the ledger, the machines were precise; the machines did accurate work. The Lords of the Admiralty declared themselves content. Brunel received a check for the money saved in one year: £17,093. Maudslay received £12,000 and the acclaim of the public and of the engineering fraternity and became generally regarded as one of the most important figures in the early days of precision engineering and one of the prime movers of the Industrial Revolution. The Royal Navy shipbuilding program would now go ahead as planned, and with the new squadrons and flotillas and fleets that were able to be created so swiftly, the British saw to it that the wars with France were duly ended, and to Britain's advantage.

Napoleon* was finally defeated, and was shipped off to Saint Helena in exile, traveling aboard a seventy-four-gun third-rate ship of the line, the HMS *Northumberland*, with as escort the smaller sixth-rate twenty-gun HMS *Myrmidon*. The rigging and other rope work of these two vessels were secured with about sixteen hundred wooden pulley blocks, almost all of them made in the Portsmouth Block Mills, sawed and drilled and milled with Henry Maudslay's iron engines, all operating under the supervision of ten unskilled navy contract workers.

Still, the ledger had two sides, and on the minus side, a hundred skilled Portsmouth men had been thrown out of work. One imagines that over the days and weeks after they were handed their final pay and told to go, they and their families wondered just why this had happened, why it was that as the need for products de-

---

* Maudslay revered Napoleon as his "ideal hero," and collected all and any piece of art that featured him. According to James Nasmyth, an engineer colleague of great note himself, Maudslay especially admired the emperor because of the great public works (roads, canals, monumental buildings, banks, the French stock exchange) he instigated.

monstrably increased, the need for workers to construct these products began to shrink swiftly away. To this scattering of Portsmouth men, and to those who relied upon these men for security and sustenance, a sum total rather too few for any serious political consideration, the arrival of precision was not altogether welcome. It seemed to benefit those with power; it was a troubling puzzlement to those without.

There was a social consequence, a reaction, although the best known, mainly because of its intermittent and spectacular violence, took place some hundreds of miles to the north of Portsmouth and was specifically involved in another industry altogether. Luddism, as it is known today, was a short-lived backlash—it started in the northern Midlands in 1811—against the mechanization of the textile industry, with stocking frames being destroyed and mobs of masked men breaking factories to stop the production of lace and other fine fabrics. The government of the day* was spooked, and briefly intro-

---

* It was pure coincidence that the prime minister whose government introduced this act, Spencer Perceval, was assassinated some eight weeks after the law was enacted. And coincidence, too, that King George III, in whose name the law was passed, was himself declared mad and temporarily removed from office. That precision-made machines were abroad at the time, and that

duced the death penalty for anyone convicted of frame breaking; some seventy Luddites were hanged, though usually for breach of other laws against riot and criminal damage.

By 1816, the steam* had gone out of the rioters, and movement generally subsided. It never entirely died, though, and the word *Luddite* (from the movement's presumed leader, Ned Ludd) remains very much in today's lexicon, mainly as a pejorative term for anyone who resists the siren song of technology. That it does so serves as a reminder that, from its very beginnings,

---

some workers were made redundant as a consequence of their introduction, and that rioting briefly erupted across the kingdom at around the same time, made the beginning of the nineteenth century a time of unusual turmoil, but not turmoil that can be blamed on the new technologies. The prime minister's assassin, for example, had a personal grudge based on a debt incurred in Russia. He was hanged for his crime, Perceval being the only British premier ever to have been assassinated.

* This figurative use of the word *steam* would not enter the language until ten years later, when the twenty-three-year-old Benjamin Disraeli included it in his first novel, *Vivian Grey*. That it was employed in literary inventions of the day is a reminder of its literal use in the still-youthful Industrial Revolution, of which Disraeli can fairly be said to have been a beneficiary, though he turned to writing to earn money, which he lost, disastrously, by investing in South American railways.

the world of precision-based engineering had social implications that were neither necessarily accepted nor welcomed by all. It had its critics and its Cassandras then; it has them still today, as we shall see.

Henry Maudslay was by no means done with inventing. Once his forty-three block-making machines were all thrumming along merrily down in Portsmouth, once his contract with the navy had been completed, once his reputation ("the creator of the industrial age") was secure, he came up with two further contributions to the universe of intricacy and perfection. One of them was a concept, the other a device. Both are essentials, even at this remove of two centuries, the concept most especially so.

It involves the notion of flatness. It involves the notion that a surface may be created that is, as the *Oxford English Dictionary* has it, "without curvature, indentation or protuberance." It involves the creation of a base from which all precise measurement and manufacture can be originated. For, as Maudslay realized, a machine tool can make an accurate machine only if the surface on which the tool is mounted is perfectly flat, is perfectly plane, exactly level, its geometry entirely exact.

An engineer's need for a standard plane surface is much the same as a navigator's need for a precise timekeeper, as John Harrison's, or a surveyor's need for a precise meridian, such as that drawn in Ohio in 1786 to start the proper mapping of the central United States. The more prosaic matter of the making of a perfectly flat surface, a critical part of the machine-made world, required only a little ingenuity and a sudden leap of intuition—both these gifts combining in the late eighteenth century in the workshop of Henry Maudslay.

The process is simplicity itself, and the logic behind it flawless. The *Oxford English Dictionary* illustrates it nicely with a quotation from the James Smith classic *Panorama of Science and Art*, first published in 1815, that "to grind one surface perfectly flat, it is . . . necessary to grind three at the same time." While it has to be assumed that this basic principle had been known for

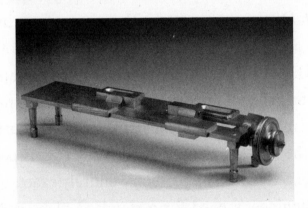

So accurate was Henry Maudslay's bench micrometer that it was nicknamed "the Lord Chancellor," as no one would dare have argued with it.

centuries, it is commonly believed that Henry Maudslay was the first to put it into practice, and create thereby an engineering standard that exists to this day.

Three is the crucial number. You can take two steel plates and grind them and smooth them to what is believed to be perfect flatness—and then, by smearing each with a colored paste and rubbing the two surfaces together and seeing where the color rubs off and where it doesn't, as at a dentist's, an engineer can compare the flatness of one plate with that of the other. Yet this is a less than wholly useful comparison—there is no guarantee that they will both be perfectly flat, because the errors in one plate can be accommodated by errors in the other. Let us say that one plate is slightly convex, that it bulges out by a millimeter or so in its middle. It may well be that the other plate is concave in just the same place, and that the two plates then fit together neatly—giving the impression that the flatness of one is the same as the flatness of the other. Only by testing both these planes against a third, and by performing more grinding and planing and smoothing to remove all the high spots, can absolute flatness (with the kind of near-magical properties displayed by my father's gauge blocks) be certain.

And then there was the measuring machine, the micrometer. Henry Maudslay is generally also credited with making the first of this kind of instrument, most particularly one that had the look and feel of a modern device. In fairness, it must be said that a seventeenth-century astronomer, William Gascoigne, had already built a very different-looking instrument that did much the same thing. He had embedded a pair of calipers in the eyeglass of a telescope. With a fine-threaded screw, the user was able to close the needles around each side of the image of the celestial body (the moon, most often) as it appeared in the eyepiece. A quick calculation, involving the pitch of the screw in inches, the number of turns needed for the caliper to fully enclose the object, and the exact focal length of the telescope lens, would enable the viewer to work out the "size" of the moon in seconds of arc.

A bench micrometer, on the other hand, would measure the actual dimension of a physical object—which was exactly what Maudslay and his colleagues would need to do, time and again. They needed to be sure the components of the machines they were constructing would all fit together, would be made with exact tolerances, would be precise for each machine and accurate to the design standard.

As with Gascoigne's invention of a century before, the bench micrometer's measurement was based on the use of a long and skillfully made screw. It employed the basic principle of a lathe, except that instead of having a slide rest with cutting or boring tools mounted upon it, there would be two perfectly flat blocks, one attached to the headstock, the other to the tailstock, and with the gap between them opened or closed with a turn of the leadscrew.

And the width of that gap, and of any object that fitted snugly between the two flat blocks, could be measured—the more precisely if the leadscrew was itself made with consistency along its length, and the more accurately if the leadscrew was very finely cut and could advance the blocks toward one another slowly, in the tiniest increments of measurable movement.

Maudslay tested his own five-foot brass screw with his new micrometer and found it wanting: in some places, it had fifty threads to the inch; in others, fifty-one; elsewhere, forty-nine. Overall, the variations canceled one another out, and so it was useful as a leadscrew, but because Maudslay was so obsessive a perfectionist, he cut and recut it scores of times until, finally, it was deemed to be wholly without error, good and consistent all along its massive length.

The micrometer that performed all these measurements turned out to be so accurate and consistent that someone—Maudslay himself, perhaps, or one of his small army of employees—gave it a name: the Lord

Chancellor. It was pure nineteenth-century drollery: no one would ever dare argue with or challenge the Lord Chancellor. It was a drily amusing way to suggest that Maudslay's was the last word in precision: this invention of his could measure down to one one-thousandth of an inch and, according to some, maybe even one ten-thousandth of an inch: to a tolerance of 0.0001.

In fact, with the device's newly consistent leadscrew sporting one hundred threads per inch, numbers hitherto undreamed of could be achieved. Indeed, according to the ever-enthusiastic colleague and engineer-writer James Nasmyth, who so worshipped Maudslay that he eventually wrote a rather too admiring biography, the fabled micrometer could probably measure with accuracy down to one one-millionth of an inch. This was a bit of a stretch. A more dispassionate analysis performed much later by the Science Museum in London goes no further than the claim of one ten-thousandth.

And this was only 1805. Things made and measured were only going to become more precise in the years ahead, and they would do so to a degree that Maudslay (for whom an abstraction, *the ideal of precision*, was perhaps the greatest of his inventions) and his colleagues could never have imagined. Yet there was some hesitancy. A short-lived hostility to machines—which is at least a part of what the Luddite movement represented, a mood of suspicion, of skepticism—briefly gave pause to some engineers and their customers.

And then there was that other familiar human failing, greed. It was greed that in the early part of the nineteenth century played some havoc with precision's halting beginnings across the water, to where this story now is transferred, in America.

Chapter 3

(Tolerance: 0.000 01)

A Gun in Every Home, a Clock in Every Cabin

To-day we have naming of parts. Yesterday,
We had daily cleaning. And to-morrow morning,
We shall have what to do after firing. But to-day,
To-day we have naming of parts.

—HENRY REED, "Naming of Parts" (1942)

He was a soldier, his name unknown or long forgotten, a lowly young volunteer in Joseph Sterrett's Fifth Baltimore Regiment. It was August 24, 1814, and I imagine the youngster was probably sweating heavily, his secondhand wool uniform patched and ill fitting and hardly suitable for the blazing late-summer sun.

He was waiting for the fighting to begin, for battle to be joined. He was hiding behind a tumbled stone wall outside a cornfield, not entirely certain where he was, though his sergeant had suggested he was in a small port city named Bladensburg, connected to the sea by a branch of the Potomac that led into the Chesapeake Bay. British forces, the word went, had landed there from ships and were now rapidly advancing from the east. Washington, the capital of his country, a country now not even forty years old as an independent nation, was eight miles to the west behind him, and he was part of a force of six thousand that had been deployed to protect it. Whispers along the line held that President James Madison himself was on the Bladensburg battlefield, determined to make sure the Britons were made to run back to their vessels and flee for their lives.

The young man doubted he would be of much use in the coming battle, for he had no gun—not a gun that worked, anyway. His musket, a new-enough Springfield 1795 model, had a broken trigger. He had fractured it, cracked the guard, and so ruined the trigger during a previous battle, an earlier skirmish of what they were starting to call the War of 1812.

In all other ways he was well enough equipped. He

had an ample supply of black powder paper cartridges, a pouch full of roundball ammunition. But the regimental armorer had told him it would be at least three days before they could forge a new trigger for him, and that he had best do all he could with his bayonet, which he had sharpened that very night, before the sun rose. Otherwise, the armorer had said with a grin, just hit the enemy hard with the gun's oakwood stock—it should give him a black eye at the very least.

That turned out not to be at all funny. The British were close by, on the left bank of the East Branch of the Potomac, when their artillery opened up later that morning, first with a deafening volley of Congreve rockets, a terrifying technique they had learned from fighting in India. It was at that moment, as massive divots of torn earth and stones clattered down around him, that the young man decided his life was more valuable than the winning of this particular battle, and that if the army couldn't be bothered to fix his musket, then he was going to run. So he turned and plunged into the high corn, heading back home to Baltimore.

He soon understood he was not alone. Through the stands of corn he could see at least five, ten, dozens of other men who were doing just the same, streaming away from the fight. Some he knew, young lads from Annapolis and the Washington Navy Yard and the Light

Dragoons, all of them apparently believing that the defense of Bladensburg was hopeless. He ran and ran and ran, and they ran, too, and all of them were still running when they crossed the line marking the District of Columbia, and they continued running, loping breathlessly in many cases, when, half an hour later, there rose before him some of the mighty structures of his capital, great buildings from where his country's government was dealing with the incomprehensible vastness of America.

He slowed to a walk. He felt he was safe now. His city was not. Before the night was out, the pursuing British troops had sacked it, more or less entirely. He found out later that the British told some of the city folk they were acting so cruelly because American forces some weeks before had had the temerity to wreck and damage buildings in the city of York, in Upper Canada. So here they burned out of revenge. They torched the half-built Capitol. They gutted the Library of Congress, and its three thousand books, and they sacked the House of Representatives. British officers dined that evening on the food Madison had been planning to eat at his Presidential Mansion, and then, after wreaking that domestic indignity, they burned his house down, too, until a ferocious rainstorm—some say a tornado—blew in and doused the flames.

The date, August 24, 1814, would be remembered for centuries to come. The Battle of Bladensburg, the last stand before the Burning of Washington and Burning of the White House, that most potent of incendiary symbols, had been one of the most infamous routs in all American history, a shameful and sorry episode indeed. The imagined account of this one soldier at war was typical of what happened that day, with battle lines being broken and troops running away in panic before the advancing enemy.

There were many reasons for the defeat, and they would be debated by clubbable old soldiers for many years. Inept leadership, ill-preparedness, insufficient numbers—the usual excuses for substantial loss have all been offered down the years. Yet one, a most notorious shortcoming of the American forces (who, after all, had fought little in the years since the War of Independence), was that the muskets with which their infantrymen had been equipped were notoriously unreliable. More important, when they failed, they were fiendishly difficult to repair.

When any part of a gun failed, another part had to be handmade by an army blacksmith, a process that, with an inevitable backlog caused by other failures, could take days. As a soldier, you then went into battle without an effective gun, or waited for someone to die and took his, or did your impotent best with your bayonet, or else, as the young man of Sterrett's regiment did, you ran.

The problem with gun supply was twofold. The U.S. Army's standard long gun of the time was a smooth-bored flintlock musket based on a model first built in France and known as the Charleville. The first of these weapons had been imported into the newly independent United States directly from France; they were then manufactured by agreement at the newly built U.S. government armory in Springfield, Massachusetts. Both models had worked adequately, though all flintlocks had misfiring problems and suffered all the simple physical shortcomings that afflicted handmade weapons that were pressed into continuous service—they overheated; their barrels became clogged with powder residue; or the metal parts broke, snapped, got bent, unscrewed, or were simply lost.

This led to the second problem—because once a gun had been physically damaged in some way, the entire weapon had to be returned to its maker or to a competent gunsmith to be remade or else replaced. It was not possible, incredible though this might seem at the remove of a quarter millennium, simply to identify the broken part and replace it with another from the armory stores.

No one had ever thought to make a gun from component parts that were each so precisely constructed that they were identical one with another. Had this step been taken, a broken part could have been replaced, swapped for another, because thanks to the precision of its making, it would have been *interchangeable*. Break a trigger in battle, and all one would have to do was fall back and get the armorer at the rear of the line to reach into his tin box marked "Triggers" and get another, ease it into place, secure it, and be back on the firing line as a fully armed and effective infantryman within minutes.

Yet no one had thought of such a thing—except that they had. Thirty years before the humiliating debacle at Bladensburg, a new manufacturing process had been created that, had it been in operation in the United States in 1814, might well have staved off a defeat occasioned by the failure of the soldiers' guns. The new thinking about the principles of gun making, thinking that, if put into practice, might perhaps have kept Washington from being put to the torch, began not in Washington, nor in the two federal armories at Springfield and down at Harpers Ferry, Virginia, nor in most of one of the stripling gun-making factories that had sprung up during and immediately after the Revolutionary War. The idea was actually born three thousand miles away, in Paris.

Back in the late eighteenth century, no one spoke about "the dark side." The phrase is modern, too new for the *OED*. In almost all the interviews for this book, about the ultrahigh-precision instruments, devices, and experiments that indicate where the precision that originates here is likely to be going, engineers and scientists referred frequently, and usually obliquely, to what "the dark side" might be doing. Once in a while, I would meet someone who admitted to having security clearance, and would thus in theory be able to discuss in greater detail what this experiment was leading to, how this device might be constructed, what the future of such-and-such a project might be—but he would invariably grin and say that, no, he couldn't discuss what "the dark side" was doing.

"The dark side" is the American military, and in terms of new weaponry or research into the unimaginably precise, that tends to mean the U.S. Air Force. Area 51 is the dark side. DARPA is the dark side. The NSA is the dark side. The role of the dark side in this story is immense, but in today's world, it is mainly to be only alluded to.

Lewis Mumford, the historian and philosopher of technology, was one of the earliest to recognize the major role played by the military in the advancement of technology, in the dissemination of precision-based standardization, in the making of innumerable copies of the same and usually deadly thing, all iterations of which must be identical to the tiniest measure, in nanometers or better. The stories that follow, in which standardization and precision-based manufacturing are shown to become crucial ambitions of armies on both sides of the Atlantic, serve both to confirm Mumford's prescience and to underline the role that the military plays in the evolution of precision. The examples from the early days of the science are of course far from secret; those from today, and that might otherwise be described in full to illustrate today's very much more precise and precision-obsessed world, are among the most secure and confidential topics of research on the planet—kept in permanent shadow, as the dark side necessarily has to be.

It was in the French capital in 1785 that the idea of producing interchangeable parts for guns was first properly realized, and the precision manufacturing processes that allowed for it were ordered to be first put into operation. Still, it is reasonable to ask why, if the process was dreamed up in 1785, was it not being applied to the American musketry in official use in 1814, twenty-nine years later? Men were running, battles were being lost, great cities were being burned—and in part because the army's guns were not being made as they should have been made. There is an answer, and it is not a pretty one.

Two little-remembered Frenchmen got the honor of first introducing the system that, had it been implemented in time and implemented properly, would have given America the guns it should have had. The first, the less familiar of the pair, despite the evidently superior nature of his name, was Jean-Baptiste Vaquette de Gribeauval, a wellborn and amply connected figure who specialized in designing cannons for the French artillery. He supposedly came up with a scheme, in 1776, for boring out cannons using almost exactly the same technique that John Wilkinson had invented across in England, that of moving a rotating drill into a solid cannon-size and cannon-shaped slug of iron. Wilkinson had patented his precisely similar system two years earlier, in 1774, but nonetheless, the French system, the *système Gribeauval*, as it came to be known for the next three decades, long dominated French artillery making. It gave the French armies access to a range of highly efficient and lightweight, but manifestly not entirely originally conceived, field pieces.* (Gribeauval did employ what were called go and no-go gauges as a means of ensuring that cannonballs fitted properly inside his cannons, but this was hardly revolutionary engineering, and it had been around in principle for five centuries.)

The second figure, the man who did the most to bring the system of interchangeable parts to the making of guns, and whose technique was, unlike Gribeauval's, unchallengeable, was Honoré Blanc. He was not a soldier but a gunsmith, and during his apprenticeship he became well aware of the Gribeauval system. He decided early in his career that he could bring a similar standardization to the flintlock musket, for the benefit of the man on the battlefield.

Yet there was a difference. A cannon was big and heavy and crude—a gunner simply touched his linstock, with its attached lighted match, to the vent, and the cannon fired—and so such parts as there were proved easily

---

* Anglo-French rivalry has been a constant for centuries, and extends into the world of war fighting just as it does into cuisine and car manufacture. Soldierly distaste for Gribeauval's purloining of John Wilkinson's work is matched by French irritation that the word *shrapnel* derives from Sir Henry Shrapnel's invention of this most deadly of weapons, a shell that hurls lethal metal debris around upon detonation. The British Sir Henry didn't invent it; a Frenchman named Bernard Forest de Bélidor did, assisted in the field by the aforementioned M. de Gribeauval.

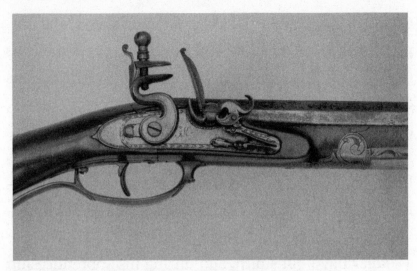

The many component parts of the flintlock on a late eighteenth-century rifle were each made by hand, and had to be filed to fit.

amenable to standardization. With the flintlock, however, the lock (that part of a musket that delivered the spark that exploded the priming powder that ignited the main charge and drove the ball down the barrel) was a fairly delicate and complex piece of engineering, made of many oddly shaped parts and liable to all kinds of failure. To the uninitiated, the names of the bits and pieces of a flintlock alone are bewildering: a lock has parts that are variously known as the bridle, the sear, the frizzle, the pan, and any number of springs and screws and bolts and plates as well as, of course, the spark-producing (when struck by the aforementioned metal frizzle) piece of flint. To render the lock into a standard piece of military equipment, with all its parts made exactly the same for each lock, was going to be a tall order.

Cost, rather than the well-being of the infantryman or the conduct of the battle, was the prime motive. The French government declared in the mid-1780s that the country's gunsmiths were charging too much for their craftsmanship, and demanded they improve their manufacturing process or lower their prices. The smiths not unnaturally balked at the impertinence of the suggestion, and promptly tried selling their products to the new armories and gun makers across the Atlantic in America, a move that alarmed the French government, as it imagined it might well run out of weaponry as a result.

It was at this point that Honoré Blanc entered the picture, taking a civilian job as the army's quality-control inspector. His brother gunsmiths expressed their dismay over the fact that one of their number was going over to the other side, was a poacher turning gamekeeper. Blanc dismissed the criticism and got on with his job, his own motivation being the welfare of the soldier out in the field rather than allowing the government to cut costs. He was greatly influenced by M. de Gribeauval, and decided he could ape his system of standardization, ensuring that all the component parts of a flintlock be made as *exact and faithful copies of one perfectly made master.*

This master he made himself, carefully and with great precision, and with all the specifications laid down as precisely as possible (using the arcane system of the Ancien Régime, which still employed dimensional measures such as the *pointe*, the *ligne*, and the *pouce*) to tolerances of about what today we would recognize as 0.02 millimeters. He then made a series of jigs and gauges to ensure that all the locks made subsequently were faithful to this first perfect master, by the judicious use of files and such lathes as were available. The gunsmiths hired by Blanc to perform this task—by hand, still—made each lock exactly as the original. Providing that they did so, exactly, all the pieces would then fit perfectly together, and the whole assembled lock would fit equally perfectly into each completed weapon.

Yet only a small number of gunsmiths were willing to work under these stringent new conditions. Most balked. Making guns simply by copying parts reduced the value of the gunsmith's craftsmanship to near insignificance, they argued. Unskilled drones could do their work instead. By arguing this, the French smiths were voicing much the same complaints as the Luddites had grumbled over in England: that precision was stripping their skills of worth. This argument would be heard many times in the future as the steady march of precision engineering advanced across Europe, the

Americas, the world. The kind of mutinous sentiments heard in the English Midlands half a century before were now being muttered in northern France, as precision started to become an international phenomenon, its consequences rippling into the beyond.

Such was the hostility in France to Honoré Blanc, in fact, that the government had to offer him protection, and so sequestered him and his small but faithful crew of precision gun makers in the basement dungeons of the great Château de Vincennes, east of Paris. At the time, the great structure (much of it still standing, and much visited) was in use as a prison: Diderot had been incarcerated there, and the Marquis de Sade. In the relative peace of what would, within thirty years, become one of postrevolutionary France's greatest arsenals, Blanc and his team worked away producing his locks, all of them supposedly identical. Blanc made all the necessary tools and jigs to help in his efforts—according to one source, hardening the metal pieces by burying them for weeks in the copious leavings of manure from the castle stables.

By July of 1785, Blanc was ready to offer a demonstration. He sent out invitations to the capital's nabobs and military flag officers and to his still-hostile colleague gunsmiths, to show them what he had achieved. Many officials came, but few of the smiths, who were still seething. Yet one person of great future significance did present himself at the donjon's fortified gates: the minister to France of the United States of America, Thomas Jefferson.

Jefferson had arrived in France the year before, to work as official emissary of the new American government alongside Benjamin Franklin and John Adams. By chance, both these men left Paris that July (Adams for

Thomas Jefferson, while U.S. minister to France, observed the early work on creating interchangeable parts for flintlock muskets, and told his superiors in Washington that American smiths should follow the French practice.

London, Franklin for Washington), leaving the intellectually curious and polymathic Jefferson alone in the ferment of prerevolutionary France. A demonstration of something scientific, with possible application for his own fledgling arms industry across the ocean, sounded like an ideal way to spend a hot Friday afternoon. Besides, it was pleasantly cool down in the château's dungeons, while up above in the Paris of July 8, 1785, it sweltered.

Honoré Blanc had arranged before him a collection of fifty locks, each gleaming in such daylight as filtered through the slit windows. Once everyone was settled on the bleachers, with onlookers paying close attention, he quickly disassembled half of them, throwing the various components of the twenty-five randomly selected locks into trays: twenty-five frizzle springs here, twenty-five faceplates there, twenty-five bridles there, twenty-five pans in another box. He shook each box so that the pieces were as disarranged as possible—and then, with a calm and an aplomb born of his supreme confidence in his method, he quickly reassembled out of this confusion of components twenty-five brand-new musket locks.

Each one of these was made of parts that had never been joined together before—but it made no difference. Everything fitted to everything, for the simple reason that with the great precision of its making, and its faithful adherence to the dimensions of the master lock, each part was identical to each other. The parts were all, in other words, exactly interchangeable.

The French officials were at first vastly impressed. The army set Blanc up in an officially sponsored workshop, he began producing inexpensive flintlock parts for the military and profits for himself, and for four further years all seemed fine. Then came 1789 and the unholy trinity of the Revolution, Gribeauval's death, and the Terror. The château was stormed, and Blanc's workshop was sacked by the rioters. His sponsor was suddenly no longer there to protect him, and there was a fast-growing, eventually fanatical, opposition among the sansculottes toward mechanization, toward efficiencies that favored the middle classes, toward techniques that put the honest work of artisans and craftsmen to disadvantage. By the turn of the century, the idea of interchangeable parts had withered and died in France—and some say to this day that the survival of craftsmanship and the reluctance entirely to embrace the modern has helped preserve the reputation of France as something of a haven for the romantic delight of the Old Ways.

In America, though, the reaction was very different, and all thanks to the prescient eye of Thomas Jefferson. The first time he described what he had seen was on

August 30, in a long letter to John Jay, the then–secretary of foreign affairs. He began with the customary flourish of logistical explanation regarding the route by which his last letter had reached Jay, an inconvenience unknown today with postal services being such a commonplace.

I had the honor of writing to you on the 14th. inst. by a Mr. Cannon of Connecticut who was to sail in the packet. Since that date yours of July 13 is come to hand. The times for the sailing of the packets being somewhat deranged, I avail myself of a conveiance [sic] of the present by the Mr. Fitzhughs of Virginia who expect to land at Philadelphia . . .

. . . An improvement is made here in the construction of the musket which it may be interesting to Congress to know, should they at any time propose to procure any. It consists in the making every part of them so exactly alike that what belongs to any one, may be used for every other musket in the magazine. The government here has examined and approved the method, and is establishing a large manufactory for the purpose. As yet the inventor [Blanc] has only completed the lock of the musket on this plan. He will proceed immediately to have the barrel, stock, and their parts executed in the same way. Supposing it might be useful to the

U.S., I went to the workman, he presented me the parts of 50 locks taken to pieces and arranged in compartments. I put several together myself taking pieces at hazard as they came to hand, and they fitted in the most perfect manner. The advantages of this, when arms need repair, are evident. He effects it by tools of his own contrivance which at the same time abridge the work so that he thinks he shall be able to furnish the musket two livres cheaper than the common price. But it will be two or three years before he will be able to furnish any quantity. I mention it now, as it may have influence on the plan for furnishing our magazines with this arm.

Jefferson was indeed seriously impressed with Blanc's system, and wrote further to friends and colleagues back in Washington, and in Virginia several times, to underline his belief that American gunsmiths should be encouraged to adopt the new French system. And in due course, the makers began to get the message, most especially in New England, where most gunsmiths were to be found.* If skepticism lingered back in Europe, America

---

* New England had more than its fair share of gunsmiths, principally because it was the first part of the continent to be thickly settled by colonists, and because there was ample water, and waterfalls, to provide power for such machinery as was needed to

proved herself, quite literally, to have the mind-set of the New World, any reluctance being swiftly dispelled by the U.S. government's decision to place enormous orders for new muskets, so long as their parts were, in line with Jefferson's thinking, interchangeable.

Two firms of private gunsmiths led the bidding for this government contract to make the first batch of muskets: ten thousand by one account, fifteen thousand by others. The winner of the contract, which meant an immediate cash payment of the not insignificant sum of five thousand dollars, was one Eli Whitney, of Massachusetts.

Whitney remains a man of great fame, still known to most in America today as he has been for two centuries. His face appears on a postage stamp. He is part of the educational curriculum. He ranks alongside inventors and businessmen—Edison, Ford, John D. Rockefeller. To any schoolchild today, his name means just one thing:

---

work primitive lathes and turning devices. Although modeled on European guns, New England weapons were often made with longer-than-usual barrels, a feature that derived from the colonists' trade with the local Indians. The main trade goods the Indians could offer were beaver skins, and it became customary for traders to exchange a musket for a pile of beaver pelts as tall as the gun was long. (One of the oldest private gun-making firms that made such weapons was that of the Robbins and Lawrence Company in Windsor, Vermont, its buildings finely preserved and lately turned into the American Precision Museum.)

the cotton gin. This New Englander, at the age of just twenty-nine, had invented the device that removed the seeds from cotton bolls, and thus made the harvesting of cotton the foundation of a highly profitable Southern states economy—but only if slaves were used to perform the work, an important caveat.

To any informed engineer, however, the name Eli Whitney signifies something very different: confidence man, trickster, fraud, charlatan. And his alleged charlatanry derives almost wholly from his association with the gun trade, with precision manufacturing, and with the promise of being able to deliver weapons assembled from interchangeable parts. "I am persuaded," he declared with a flourish of elaborate solemnity in his bid to make a cache of guns for the U.S. government, "to make the same parts of different guns, as the lock for example, as much like each other as the successive impressions of a copperplate engraving."

It was the utmost piffle. When Whitney won the commission and signed the government contract in 1798, he knew nothing about muskets and even less about their components: he won the order largely because of his Yale connections and the old alumni network that, even then, flourished in the corridors of power in Washington, DC. Once he had the contract in hand, he put up a small factory outside New Haven and promptly claimed to be manufacturing muskets there, weapons based, as were all smooth-bore American guns of the time, on the French Charleville design. He took an unconscionable time to produce any weapons, however. The contract specified a delivery of at least some of the muskets by 1800, but there were only a handful of finished guns, and all Whitney could offer as a salve by that due date was a demonstration of the quality, as he claimed, of the guns that his new factory was now notionally in the process of making.

Whitney performed what is seen as his notorious demonstration in January 1801—a supposed confidence-building exercise, it would be called today—before a distinguished audience that included the then-president, John Adams, and his vice president, soon to become president, Thomas Jefferson, the man who had started the ball rolling fifteen years before. There were also dozens of congressmen and soldiers and senior bureaucrats, all men who needed to be convinced that public treasure was going to be expended on what would be a truly worthwhile venture. They had been told they were there to witness Whitney demonstrating, with the use of a single screwdriver, how his musket locks were properly interchangeable.

Everyone in the room was ready to believe him, Whitney's cotton-gin-based reputation having long preceded him. It seemed to be of no great moment to anyone in the room, however, that the man didn't even bother to disassemble the locks he had on show. Instead, he merely took a number of finished muskets, used his screwdriver to detach the locks from their wooden gunstocks, then slipped them whole into slots on other gunstocks, and so made it appear to the guileless visitors as though his parts were, as promised, truly interchangeable.

He explained as he went along what he was doing, and not even Jefferson, who had seen Blanc's demonstration at Vincennes in 1785 and might have had sufficient knowledge to splutter, "Hold on a minute!" had the temerity to challenge him, to express even the smallest measure of skepticism. Quite the reverse: the president-elect bought Whitney's explanation in its entirety, and wrote enthusiastically to the then-governor of Virginia, saying that Whitney had "invented moulds and machines for making all the pieces of his locks so exactly equal, that take 100 locks to pieces and mingle their parts, and the hundred locks may be put together as well by taking the first pieces that comes to hand."

The truth is Jefferson had been hoodwinked, as had everyone else present that day. For there had been no molds, no machines for making all the parts "so exactly equal." Whitney's new-made factory, powered by water, not yet by steam (even though engines were readily available), had neither the tools nor the capacity to make precision-engineered pieces. Realizing this, he had instead hired a clutch of artisans, craftsmen, and told them to make the flintlock components with their own files and saws and polishers, and make them one by one, by hand—and not necessarily all the same, either, for the way he had planned his show did not allow for anyone to inspect the locks themselves, only that they fitted into the stocks.

So there was no new technique. Everything had been done the old-fashioned way, but with the demonstration's ringmaster, the master of ceremonies, working to convince all in the room that they had just seen a remarkable and revolutionary manufacturing process, live and in the flesh. Nothing about the display was genuine: no lock had to be taken apart, and even the gunstocks were preselected to make absolutely certain the slot in each of them was large enough to accommodate whichever of the ten locks was chosen as a replacement.

Muskets made by Whitney survive in collections to this day, and they reveal the sorry story: that the promise of precision, with its reward of easy money, led to cunning and corruption. None of the surviving weapons is well made; nor do its locks show any indication of exacting similarity. They might well fit into the stocks, but their parts would not fit into one another.

The demonstration worked, though. The sheer flamboyance of Whitney's spectacle did end up convincing the government to hand him a further sum of much-needed money, even though those attending the demonstration had, to a man, been duped. Whitney was a fraud, and the fact that it took a further eight years before his guns were delivered suggests that, in the end, those who handed over the cash got all they deserved.

True credit for taking Honoré Blanc's French system and translating it into the American way of precision-based manufacturing actually belongs to three lesser-known figures: Simeon North and John Hall, gun makers, and to Thomas Blanchard, who could do remarkably replicable things with wood. North had his smithy not twenty-five miles away from Whitney's factory in Middletown, Connecticut. John Hancock Hall was from farther away, in southern Maine, and he had made something of a fortune running first a tannery and later a series of cabinetmaking and boatbuilding wood shops. Guns were a sideline, a hobby—until, in 1811, when he filed applications for a patent for an entirely new kind of weapon: a gun of his own invention

Stacked guns in the so-called musket organ at the U.S. government's Springfield Armory in Massachusetts, where the French system of making interchangeable parts revolutionized manufacturing.

and design, a single-shot rifled weapon that could be loaded through the breech rather than, as with muskets, down through the barrel.

In time, both men, North and Hall, won government contracts for producing guns—North for horse pistols in Connecticut; Hall for his new breech-loading weapons up in Portland and then, later on, down at one of the two newly established federal armories in Harpers Ferry, Virginia. (The other was at Springfield, Massachusetts.) The rather more significant breakthrough made by both men—by all three men, in fact, though Blanchard's was in a subsidiary and less focused role—was that, for the first time, they each used machines to make their gun components. This was a major change, and in making it, the men ensured, rather than simply hoped, that what was made was near perfect and true and precise, every time.

Those who had initially planned for interchangeability, Blanc and Gribeauval in France and those in the U.S. government who had impenitently asked Eli Whitney to do as he had promised, did so by employing workers to hand-make their components and to keep them true to a master example of each piece. They achieved good results by making jigs, gauges, and master models. The workingmen they hired to perform the various tasks, all the while complaining that their time-honored skills were going to waste, had to create new pieces by using the jigs, then measure the pieces using the gauges, and finally compare their dimensions with those of the masters, confirming thereby that they were exact copies, and thus producing de facto interchangeability.

But humans are fallible, however legendary their craftsmanship. The hand of the man who shapes, the eye of the man who smooths, the mind of the man whose claims to inerrancy—all suggest he instinctively knows when something is *right*, yet all can and will eventually misjudge, make mistakes, fall afoul of fatigue. Machines, on the other hand, if properly set up and not yet worn out, are well-nigh incapable of error. Those ma-

chines that can perform the kind of tricky work hitherto reserved for skilled artisans (such as the abundance
of machines made by Henry Maudslay for the naval
pulley block factory in Portsmouth) can almost guarantee perfection and consistency in their production. The
machine offers what one historian has called "the workmanship of certainty . . . in which the result is predeter-
mined and unalterable once production begins."

And what North and Hall were able to do, independently, was to create machine tools that offered just that
degree of certainty. Simeon North up in Middletown
made one of America's first metal-milling machines,
replacing at a stroke the tedious handiwork of filing and
checking, filing and checking, and instead putting a
belt-driven cutting tool to work milling away the super-
fluous metal, while a mixture of oil and water kept the
cutter and the workpiece cool as it was being reduced,
smoothed, and shaped.

John Hall, working five hundred miles to the south,
in a government-gifted metal shop right beside the
Harpers Ferry arsenal, then improved upon this milling machine,★ and built a series of what were called

---

★ Improvements made a long while ago can seem mundane and
trivial with the benefit of contemporary sophistication, but were
critical in the evolution of precision engineering. John Hall's improvements are very much of this kind: he tinkered with the

drop-forges, which he sited upstream, as it were, of the milling devices in his workshop. A long piece of red-hot iron, soft and pliable, was forged between hard-tempered metal dies, one of them static, the other one lifted and repeatedly and heavily dropped onto the other until the piece between them (by now drop-forged) was roughly shaped—into a gun barrel, say—and then handed over to the men working the milling machine.

Employing a variety of differently designed cutting tools fixed to the milling head, these men would mill away excess iron from the forged rod in order to shape and trim and turn it into a tube of iron that could then be rifled and made into a useful central part of a working gun. At every stage of the work, from the forging of the barrel to the turning of the rifling and the shaping of the barrel, John Hall's gauges were set to work—he employed no fewer than sixty-three of them, more than any engineer before him, to ensure as best he could that every part of every gun was exactly the same as every

---

means of ejecting a workpiece from the milling machine, and so prevented the die's temperature from changing dramatically during the process and thus risk losing its temper. He also designed so-called fixtures, the devices that hold a workpiece absolutely secure during milling, further ensuring that his milling cuts were made with all needed precision, an essential for guaranteeing the fit of the pieces.

other—and that all were made to far stricter tolerances than hitherto: for a lock merely to work required a tolerance of maybe a fifth of a millimeter; to ensure that it not only worked but was infinitely interchangeable, he needed to have the pieces machined to a *fiftieth* of a millimeter. And once the barrel, made with such a strict adherence to rules and numbers, had been shaped and checked and checked again, it remained necessary only to have the flintlock attached to it and the whole inserted into the wooden stock—which is where the last member of this holy trinity of early American precision engineers, Thomas Blanchard, comes in.

In 1817, in his hometown of Springfield, Massachusetts, Blanchard invented a lathe that made lasts for shoes. It was a stroke of inventive genius: he simply placed a metal template of a shoe in his machine and, using a pantograph connected to a series of blades, attached the template to the shapeless hunk of ash, a last-to-be that was fixed in the path of a series of sharp knives. Turn the template, trace its outline with the pantograph rods, and let the other ends of the pantograph in turn press the blades against the timber—and presto! In ninety seconds or less, an exact copy of the template would be there, in freshly carved wood, ready to be taken from the machine and sent off to the cobbler.

One simple consequence of such a machine lives with

us today, in the matter of shoe sizes. For as Blanchard could now turn a shapeless block of wood into a foot-shaped entity of specific dimensions, and repeat the creation time and time again, so he could offer to the shoemaker lasts of different but exact sizes—one that was seven inches long, one nine, and so on. Prior to that, shoes were offered up in barrels, at random. A customer shuffled through the barrel until finding a shoe that fit, more or less comfortably. Now he simply asked for a size seven, or eleven, or five medium.

And as with shoes, so later with gunstocks. Blanchard was soon offered work at the huge and growing Springfield Armory nearby, and was asked to adapt his shoe last lathe to make the wooden parts for guns that, though necessarily more complicated than feet, had the benefit of being needed in only one size. So he made a metal model of a gunstock (an irregular form, in the same sense that a foot is structurally unique) and set it high on the lathe, connected to a pantograph as before. And turning on the rotating driver of what was described as "a strange contrivance . . . at first glance less like a lathe than some primitive piece of agricultural machinery," he commenced the process of regular gunstock manufacture, a process that survived at the armory for more than the next half century. Thomas Blanchard had cleverly patented the principle of his lathe, and a company in the nearby town of Chicopee manufactured it under license. The inventor lived on into old age, comfortably settled by a near-ceaseless fountain of royalties.

The management of the Harpers Ferry Armory was eager to try out all these new contrivances—despite its remote location, the armory was more accepting of innovation, oddly, than was the busier, bigger, older armory at Springfield, where Blanchard worked, and at which Simeon North was a regular visitor. Harpers Ferry became almost certainly the first establishment in the United States, maybe the first in the world, to employ precisional techniques and mass production to create weapons for the country's military. To do so, it employed an array of these new technologies and ideas. It used the products of Blanchard's gunstock machine; it also used John Hall's milling machine, his fixtures, and his drop-forges; and its locks were made by the process invented by Honoré Blanc and perfected by Simeon North. From iron smelted in Connecticut to finished guns smelling of linseed oil (for the ashwood stock) and machine oil (for the barrel and lock), these were the first truly mechanically produced production-line objects made anywhere—they were also American and, just as Lewis Mumford had predicted, they were guns. Also, they were machine-made in their entirety, "lock, stock, and barrel."

The newborn manufacturing community had other irons in the fire besides, and most of them of a decidedly nonbelligerent nature. One Oliver Evans was making flour-milling machinery; Isaac Singer introduced precision into the manufacturing of sewing machines; Cyrus McCormick was creating reapers, mowers, and, later, combine harvesters; and Albert Pope was making bicycles for the masses. And while the Northeast of the United States has long worn its still-surviving reputation for firearms making—the broad lowland reaches of the Connecticut River have long been known as Gun Valley, as gun makers were (and mostly still are) all here: Colt, Winchester, Smith and Wesson, Remington—it was soon to be known for other creations: for another high-precision industry had lately moved into the valley towns and cities of America at about the same time.

Those who operated the machines that were locally bent to making the small components for the region's armories (the triggers, the faceplates, the frizzle springs) found that they could with ease modify their lathes and milling machines to make small gearwheels and spindles and mainsprings, the necessary components for the intricacies of timekeeping. The region, in consequence, became famous for the production of clocks, for generations of precisely made, and occasionally accurate, plainly beautiful American timepieces.

I write this to the steady beat of a Seth Thomas thirty-day kitchen clock, made in Plymouth, Connecticut, in the 1920s. It is a thing of solid utilitarian beauty, the sort of thing the Shakers would have made if they had concerned themselves with time beyond daybreak and dusk. It is not alone: there are many other clocks scattered around this old farmhouse, most of them eight-day clocks, five of which need winding every Sunday morning, one that has as its pendulum two cylinders half-filled with liquid mercury. In the hall there is a long-case clock made in Winchester, Connecticut, which I bought for reasons of eponymy and which is a little troublesome: it is more than a century old and has wooden gearwheels, which are inconveniently susceptible to changes in the ambient temperature and humidity. The others are more or less reliable, though, and so long as I wind them with an eye to synchronicity, they all remain ticking and chiming as they should, with the exception of one in the kitchen, a former British Railways station clock that has a mind of its own and sometimes demands winding in midweek, which I find confusing.

Still, what I particularly like about old-fashioned clocks is that they may well have been made precisely (their gearwheels fashioned to tolerances of some thousandths of an inch, their springs tightenable to precisely calculated and specific torques, their pendulum bobs precisely weighed, and their pendulum sticks of exactly measured lengths), but they are often anything but accurate. And part of the pleasure of my Sunday morning ritual is correcting them all, pushing this hand a little forward, that one a minute or so backward, putting the grandfather (which gains inordinately) back by ten minutes or more.

One of the best-loved films of my childhood was *The Fallen Idol*, a genteel drawing room thriller made by Carol Reed, in which most of the drama takes place inside the French embassy in London. One scene remains in my mind: at the same moment as the details of what looks like a gruesome murder are being unpicked by a group of burly policemen, the Sunday morning clock winder makes his appearance, performing work on the embassy's elegant clocks, all ormolu and cloisonné, just as I do on my much humbler collection. Hay Petrie, a diminutive character actor from Dundee, has the role, and checks the clocks by his own pocket watch, presumably an impeccable timekeeper. My own domestic standard timekeeper is a pocket watch, too, a Ball railroad watch wound daily and which keeps to about ten seconds a week. When, every month or so, I find it necessary to reset this, I telephone the time recording from the U.S. Naval Observatory master clock, which has as its own standard a series of cesium fountain atomic clocks in a secure building in Boulder, Colorado.*

Though by Sunday breakfast all my clocks are in harmony, it takes only a day or so for them to fall slightly out of rate once more. By Wednesday, I head up to bed listening, just as Harriet Vane does when, in *Gaudy Night*, she listens to Oxford's clocks, the various iterations of midnight being chimed out in "friendly disagreement." In writing that line, Dorothy Sayers was celebrating a mild and meaningless inaccuracy from which one might well take (as I most certainly do) a considerable but inexplicable satisfaction.

To the ordinary and reasonable human, there can perhaps be too great a degree of, or reliance upon, precision, which is something the clockmakers of New England understood well. They knew that the use of

---

* Being able to connect the time in my various household clocks to the atomic clock that provides the official time for the United States of America introduces the concept of traceability, a cornerstone of precision unknown to the clockmakers and gun makers and the pulley block makers of the eighteenth and nineteenth centuries, but entirely essential today. Much is to be made of traceability in the world's metrology institutes, described in the afterword.

interchangeable parts made the manufacture of things a great deal easier than before, and that they could make their goods both quickly and, most important for customers, cheaply. They knew also that accuracy was not of supreme importance in clocks, even though such a sentiment seems to fly in the face of what a clock is intended to do.

Both precision and accuracy are crucial in the making of guns—a soldier's life depends on his weapon, on its reliability and the honesty of its making—but a clock in a family home, in an early nineteenth-century home, that is, was there more for the decorative augmentation of the kind of daily events that marked time more conventionally: the passage of cows from meadow to byre, the children's morning yearning for breakfast, the blast of the steam whistle, the peal of church bells. Clocks of the kind being made in America, necessarily very different from the kind of timekeepers John Harrison had been making for the Board of Longitude in England in the previous century, were offered as symbols of arrival into the middle class, much as were sewing machines and washing machines, also Connecticut Valley–made at around the same time.

Clocks that were cheap, repairable, moderately accurate—these were the requirements of the customer, and it was the benefit of precision-based engineering that allowed them to be made so. Perhaps we should not be as surprised as the visitor to the American West in the middle of the century who remarked that "In Kentucky, in Indiana, in Illinois, in Missouri, and in every dell in Arkansas, and in cabins where there was not a chair to sit on, there was sure to be a Connecticut clock." That was part of the triumph of a means of making that was already being called, to the envy of all industrialized nations around the world (including the British, who could still rightly lay claim to having been the pioneers of precision and perfection), the American system.

Chapter 4

(Tolerance: 0.000 000 1)

On the Verge of a
More Perfect World

All of beauty, all of use
That one fair planet can produce,
Brought from under every star,
Blown from over every main,
And mixt, as life is mixt with pain
The works of peace with works of war.

—ALFRED, LORD TENNYSON, "Ode Sung at the
Opening of the International Exhibition" (1862)

On the warm and sunny midafternoon of Monday,
July 2, 1860, in the then-leafy London village

suburb of Wimbledon, Queen Victoria performed a task many of her subjects would have thought unsuited to her dignity, improper for her sex, and inappropriate to her station. She fired a high-powered rifle, and with a single shot over a range of a near–quarter mile, she scored a near-perfect bull's-eye.

It was all a little more complicated than it sounds. Her Majesty did not simply adjust her crinoline, fling back her veil, hurl herself to the ground, and let loose at a distant target. This was the opening moment of an international contest run by Britain's National Rifle Association, of which the queen was patron, and she had been asked to inaugurate the event in an appropriate manner. There should be an opening gunshot, it was thought, and the queen should fire it. To the surprise of all, the Palace agreed—subject to certain conditions. Her Majesty was not going to lie on the royal stomach, or prostrate herself whatsoever.

Accordingly, on a crimson silk–swathed dais built near the pavilion tent where the queen would arrive from Buckingham Palace, there stood a gleaming state-of-the-art Whitworth rifle. It wasn't just standing propped up on the side; it had been firmly mounted on a stout iron stand and was pointing toward the leftmost of a line of targets that stood before a range of butts four hundred yards away across Wimbledon Common. The

Joseph Whitworth's name is memorialized today in the standard measurement of screw threads, BSW, for "British Standard Whitworth." He also designed rifles much used by the Confederate side in the U.S. Civil War.

gun was set horizontally, at a height commensurate with the queen's modest stature: mighty she might be to her subjects, but she stood just four feet eleven inches, a height of some significance, though, when someone fires a gun while standing up. A silk string with a tassel was attached firmly to the gun's trigger. The safety catch was on.

Nothing was going to be left to chance, and in consequence, Joseph Whitworth, the Manchester engineer who three years before had invented and designed this hexagonally barreled, .45-caliber high-powered weapon, was extremely nervous and concerned. Working with a team of assistants, he had spent two harried hours that afternoon adjusting his demonstration gun to bear precisely on its target. His reputation (stellar but, like all reputations, vulnerable) depended absolutely on the success of this firing. If the gun misfired, his hopes for high favor would be forever dashed. If the queen missed the target, he would be socially ostracized. And if, heaven forfend, Her Majesty's bullet accidentally hit and killed someone . . .

The hundreds in the audience waiting for the arrival of the queen didn't see it quite that way, and were most amused as Whitworth's test shots crept closer and closer to the red circle at the center of the target. "Much signalling with flags passed between the tent and the markers at the target," wrote the reporter from the London *Times*. "Then more manipulation. Then another shot, till a short time only before Her Majesty's arrival a satisfactory adjustment was arrived at."

Whitworth checked that a .45-caliber bullet was in the chamber. Finally, he set the safety catch to off.

Queen Victoria arrived on the scene shortly before the appointed hour of 4:00 p.m. Her entourage included her beloved Albert, naturally; a gaggle of young princes and princesses; and a small battalion of top-hatted court officials and prim ladies-in-waiting. Functionaries of great seniority and solemnity greeted her, then escorted her and Albert to the Rifle Tent and its silk-swathed dais. Joseph Whitworth, nervously arranging and re-arranging his tie, waited. The queen waited, too, the polished rifle beside her.

From all around the Common, church bells then began their preludes to pealing the hour. It was 4:00 p.m., on the dot, and Her Majesty, not having even seen the target but fully briefed on what she should do, reached over, grasped the tassel, and tugged gently on the silk string. Nothing happened. Maybe she pulled too lightly, so she tried again. Then she was met with slight resistance, and as briefed, she then tugged harder, a third time. This did the trick.

There was a sudden loud report—a crack!—and then a gust of black smoke from the rifle's barrel, neither of which seemed to startle the royal personages. A few seconds went by, everyone keeping silent as the royal gunshot echoed and reechoed around the fields. Then, suddenly, in the far distance, a red-and-white flag was jauntily hoisted and could be seen waving in front of the target.

A gale of wild applause and cheering immediately swept out from the loyal crowd. The queen, without either intention or challenge, had not just hit the target but had done so dead center. A small smile wafted across her face, as if she were faintly amused.

She had scored a bull's-eye. Close forensic examination showed that over the four hundred yards of travel, her bullet had deviated only an inch and three quarters in elevation and four-fifths of an inch from the direct line. She had been, or was believed to have been, both precise in her aim and accurate in her intended result.

And with that single shot, the 1860 Grand Rifle Match of Britain's National Rifle Association formally got under way, with all concerned, Joseph Whitworth most especially, happy and mightily relieved.

Queen Victoria, Prince Albert, and Joseph Whitworth had met once before, nine years prior to this encounter in Wimbledon. (Victoria and Whitworth would then meet one further time, nine years later, when she conferred on him the honor of a baronetcy, a hereditary knighthood, for services to engineering. By then, the queen wore black; her adored Albert had died in 1861.)

In midcentury Britain, there was a very real sense that the Western world was changing, and changing fast. The social revolution that had been begun by James Watt and his steam engine had by the middle of the century properly taken hold, and industrialization was affecting everyone's life, for good or for ill. Cities were swelling, villages were wilting, factories were being thrown up, mines were being sunk, railways were snaking across the landscape, docks were busy with trade, chimneys were belching smoke into previously unspoiled air, wages were being earned, trade unions were being formed, and an extraordinary popular appetite for science and technology was discernible. *Progress* was the word on everybody's lips, and the feats and possibilities of machinery were inspiring awe and apprehension.

Halfway through the nineteenth century, humankind, Western and industrializing humankind most particularly, had somehow reached a hinge point, a time for some to stop and take stock. And in London, capital of the country that, at the time, was near-universally seen as the intellectual, spiritual, and scientific center of the Western world, it was decided, and decided essentially by royal command, that it would be meet and proper to savor the moment, to show off what had been achieved in the world thus far, and to offer some thoughts on what might be coming next.

A Great Exhibition was proposed and conceived, a celebration of achievement to be entitled in full the Great Exhibition of the Works of Industry of All Nations of 1851. The French had been holding modest but fairly regular displays along these lines in Paris since

The Great Exhibition of 1851, staged in London's Hyde Park, allowed the Western world to consolidate under the enormous roof of the Crystal Palace the inventions of the Industrial Revolution to an enthralled public.

the end of the century; Berlin similarly staged a small celebration of achievement a few years later; and in London, the Society of Arts* held a competition, with prizes, dedicated to industrial design in 1845. What was planned for 1851, however, was a spectacle intended to blow all its predecessors memorably out of the water. And Joseph Whitworth, though little known beyond his particular calling, was to be one of those invited.

---

* It may be recalled that it was at the same society that Joseph Bramah, sixty years before, first encountered the complexities of locksmithing and made what he thought was an unpickable lock. It was finally picked at the 1851 exhibition.

It was Queen Victoria's imaginative consort, Prince Albert, who remains most publicly associated with the idea of staging a Great Exhibition. With a degree of foresight still admired two centuries on,* he came to recognize the time's extraordinary zeitgeist, and he wished to capture its uniqueness for one shining summertime, and present it, in a grand and spectacular manner, to his public. He wished the world to hold up a mirror to itself and see just how memorable was its history, then so busily unfolding. Moreover, so confident was he of the popular fascination with what so enthralled him that he was sure such an exhibition would in time pay for itself. Accordingly, as he painstakingly selected the members of the commission that would plan it, and as he meticulously planned who should be invited and what kind of creations should be on show, he made a single stipula-

---

* The origin of the exhibition idea properly belongs to Henry "Old King" Cole, a British civil servant of remarkable ability and breadth of knowledge who, among other achievements, designed the world's first postage stamp, the "Penny Black." Cole also began the tradition of sending Christmas cards each December (and printed his own), and under the pseudonym Felix Summerly, he won an award at the 1845 Society of Arts Exhibition for the design of a ceramic tea service. He knew Prince Albert well, and persuaded him to defy the insufferable court traditionalists by throwing his weight and influence behind this immensely ambitious project of 1851.

tion: that the exhibition be financed privately, and not from the public purse.

"We are living," Albert declared at the banquet that inaugurated the fund-raising effort, "at a period of most wonderful transition, which tends rapidly to accomplish that great end—to which all history points—the realization of the unity of all mankind. Gentlemen, the Exhibition of 1851 is to give us a true test of the point of development at which the whole of mankind has arrived in this great task and a new starting point from which all nations will be able to direct their further exertions!"

By way of making such stirring addresses, Albert soon managed to find all his money in double-quick time, and he then had a polymathic gardener named Joseph Paxton design and then throw up on the southern side of Hyde Park an enormous structure built almost entirely of glass and iron, 1,851 feet long to celebrate the year of the exhibition and 108 feet tall at its highest point such that it could accommodate three of the park's ancient and best-loved elm trees, which now needed not be felled. The Crystal Palace, as it came to be called, took only six months to build. With nearly a million square feet of glass panels, it looked like a truly fantastic greenhouse, a greater version of the hothouse that Paxton, as gardener, had built for the Duke of Devonshire's collection of lilies.

And here, for only a modest price—"The World for a Shilling" was the slogan that attracted visitors by the tens of thousands—were gathered, among myriad marvels, a collection of enormous, heavy, impressive, fully working, and frequently roaring-hot ironbound inventions that were the most up to date, the most important, and among the most visited items on show. They were machines, great big British iron machines; machines that showed, and with a certain sense of disdain, that however obsessed America might be with the cleverness of her precisely made interchangeable parts, however pleased with the consequent beginnings of mass production and, if yet some way ahead, with the makings of the assembly line, this was a moment in British history when mechanical brute power and might were the things to be displayed and deployed. For America, such display would come later. For now, this was Britain's time, and presentations of national endeavor built on a grand scale would mark the moment.

Patriotism, together with a pervasive sense of jingoism, had naturally much to do with the local popularity of these British machines. While British people of the time certainly liked to be titillated by the trivial and the amusing, of which the exhibition had plenty, it was clear also that it was through the making and use of these monumental inventions that Britain, soon to be at

her imperial apogee, at her proudest and most powerful, would continue to prosper, dominate, and rule.

For a time, at least. If there was a faint drumbeat of
doubt, Britons of the day were quite deaf to it. They
were happily fascinated by the steady march of their
constructions—the huge ships, the great guns, the soaring iron bridges, the canals, the aqueducts. The still very
new sight of steam locomotives, the best of them gleaming in their green and red and black enamel paint and
with highly polished brass, could be guaranteed to draw
crowds at any railway terminus, and the swelling number of water pumping stations and printing presses, and
the solemnly rocking iron beam engines that powered
them, never failed to capture the popular imagination.

Yet that same imagination could barely conceive of
the diverging paths on which America and Britain had
now accidentally set themselves. Nor could any see
that the British path might well lead into a technological cul-de-sac, while America's would lead, at least for
a while, toward a more open road of development and
progress. In 1851 there seemed no stopping the British
Isles, and the inventions she had on display were indeed illustrations of what was widely believed to be her
unchallengeable power, on the move and for always.

For the benefit of the visitors, the exhibits were arranged in broad classes—Class 1: Mining and Min-

eral Products; Class 2: Chemical and Pharmaceutical Products; Class 3: Substances Used as Food; Class 4: Vegetable and Animal Substances Used in Manufactures; Class 5: Machines for Direct Use including Carriages, Railway, and Marine Mechanisms; Class 6: Manufacturing Machines and Tools; Class 7: Civil Engineering and Building Contrivances; Class 8: Naval Architecture, Military Engineering, Guns, Weapons, Etc.; Class 9: Agricultural and Horticultural Machines and Implements; and so on—thirty classes in total, all rich in their variety and technological achievement.

To drill down into any one of these and to explore would be to confirm Prince Albert's view that the mid-nineteenth century was a moment of "wonderful transition." To drill down particularly into Class 6, Manufacturing Machines and Tools, is to explore, quite literally, that same transition's *cutting edge*,* most especially where it related to items that had been made with the utmost care and precision.

---

* Lexicography would amply confirm this: for though the exact meaning of the English phrase "the sharp edge of a blade that performs the cutting" had been in use since 1825, the figurative meaning, in which *cutting edge* is "the latest or most advanced stage in the development of something," first appeared in print in an American journal called *The National Era*, in July of the very same year as the Great Exhibition, 1851.

Here were the machines of the future, and the mechanicians who would make them. Messrs. Waterlow and Sons, for example, had invented an automatic envelope-making machine, which drew long queues of the curious. People would feed in a sheet of paper that, in a blink of an eye, would be cut, folded, and gummed, ready for its letter and a stamp. A company in Ipswich had come up with a steam-powered excavator for cutting through low hills to allow passage for railway lines—such a monster had never been seen before, nor ever imagined. Another company, based in Oldham in Lancashire, had brought down some fifteen cotton-spinning machines, each one of them, like all the moving devices in Class 6, being sited close to the boilers that had been built in a separate structure outside the Crystal Palace and that piped steam in to make the machinery work.

Robert Hunt, a Victorian science writer who undertook a two-volume, 948-page labor of love, *Hunt's Hand-Book to the Official Catalogues*, in which he described and critiqued every last object in the Crystal Palace, was particularly impressed by the Oldham exhibit. The "fingers of the spinners . . . with the aid of that classical instrument the domestic spinning wheel," he wrote, have at last and for all time been superseded by this machine, which has "several thousand spindles . . . in a single room, revolving with inconceivable rapidity, with no hand to urge their progress or to guide their operations, drawing out, twisting and winding up as many [as a] thousand threads with unfailing precision, indefatigable patience and strength—a scene as magical to the eye that is not familiarized to it, as the effects have been marvellous in augmenting wealth and population."

Robert Hunt did fret, somewhat. At the close of one particularly lyrical passage about a new power loom, he wrote, "Wonderful mechanical result! What are the moral results?" and repeated his concerns in a similar manner throughout his writings. But few other visitors or critics seemed to share his sentiments, or worried about the social implications. Not in Britain, anyway. The French were perhaps most aware that there might be a downside to all that "unfailing precision": the ennobled mathematician and politician Charles Dupin warned that "by superseding the labour, the country is depopulated and filled with machines," and it would be up to the politicians of the future to decide if that was progress. Clearly the good baron thought it was not progress at all—a view shared famously some twenty years later, when his fellow countryman, Gustave Doré, produced his book of engravings of London slum land, which many saw as an indictment of the New World, an amply deserved reminder of the lack of social progress that precision had somehow conspired to create.

The great majority of the thousands who came were happy to see as many examples of steam-powered mechanical perfection as were on offer. To them, the machines were just magical things—the looms, the printing presses, the railway locomotives, the trams, the marine engines (the most impressive of them made by the firm of Maudslay, Sons and Field, which forty years before had designed and created the block-making machinery for the Royal Navy, and was still going strong), and the early and the more refined Watt steam engines themselves. Some other sources of power were on show, waterwheels and windmills most especially, and there were early horse-drawn omnibuses, one with two floors and a spiral staircase mounted aft, an early version of the London double-decker bus. Yet it was steam-powered engines, with their glare of radiant fire, their thunderous sounds, the smell of hot oil, the sheer vision of power they seemed to force on those who gathered awestruck in front of them, that remained most indelibly in the mind's eye. The audience for them was obliged to stand behind protective railings, for these were dangerous engines, hurling fast-moving, highly polished bars of iron and spinning two-ton gearwheels through space, and capable of easily smashing skulls and catching limbs and sweeping whole children into their maws. These were machines that people would love, but of which they were rightly fearful, and from which they would keep their distance.

Amid all the rousing chaos was a quieter side to Class 6, a display of more static British-made machinery, the long-term importance of which was, if anything, even greater than the mesmerizing whirligigs that drew the largest crowds. And presiding over this quieter byway of the exhibition, at Stall No. 201, was the Manchester-based firm founded by the man universally then known, and still today most widely regarded, as perhaps the greatest mechanician in the world, the man who, nine years later, would bite his nails with nerves while watching Queen Victoria fire one of his rifles. "Whitworth, J & Co," the catalog reads. "Self-acting lathes, planing, slotting, drilling and boring, screwing, cutting and dividing, punching and shearing machines. Patent knitting machine. Patent screw stocks, with dies and taps. Measuring machine, and standard yard &c."

A rather unprepossessing description, to be sure. It hardly improved when Joseph Whitworth himself put in an appearance on those days when, from time to time, he came down from Manchester. He was large and bearded and oyster-eyed, rather frightening-looking—he had a face "not unlike that of baboon," according to Jane Carlyle, the wife of Scottish social commentator Thomas Carlyle—and, besides his fearsome looks, was also known for his irascibility, his unwillingness to suffer fools gladly, his domineering manner, and (on a personal level) his relentless infidelity. But the twenty-three instruments and tools he had on show during those six months in London, though they may have lacked the luster and swash of big steam engines and thousand-spindle looms, provided a road map to what would become engineering's future (and won their maker more medals than any other of the Crystal Palace exhibitors). Joseph Whitworth was an absolute champion of accuracy, an uncompromising devotee of precision, and the creator of a device, unprecedented at the time, that could truly measure to an unimaginable one-millionth of an inch. Before him there was precision; afterward, there was Whitworth-standard precision, and the Great Exhibition was where he made his reputation for it.

The big-name turn-of-the century engineers all seemed to know one another, to train one another, to be apprenticed to one another. Whitworth was very much part of this picture. His profound interest in mechanical perfection began when he was a very young man—he had been effectively orphaned after his mother died and his father took off to train as a priest—and was apprenticed to Henry Maudslay. It was during his time with Maudslay that Whitworth first became fascinated by the very particular idea of the flatness of surface plates.

As Henry Maudslay had already demonstrated, perfect flatness is a thing of the utmost importance. Its elemental significance is quite simple, is central to what one might call the philosophy of precision. A perfectly flat plate does not derive its perfection from anything else—it isn't something to be measured in relation to something else. Its dimensions are unimportant. Its shape is of no note. It is either flat or it isn't. And by being exactly flat, it can give precision to those other things that are measured against it. A ruler, a square, a gauge block—all can be set against a flat plane and declared to be true, or not; precisely made, or not.

So, perhaps not unreasonably for the two men for whom this concept was of such paramountcy, there was a small squabble over who had in fact come up first with the means of achieving it. For a while, the dispute flared. But time has now taken care of the argument. Maudslay is given his due as the originator of the notion, the discoverer of the principle: what Whitworth did was to improve and then expatiate upon this notion and give it teeth, as it were—and by doing so, he immodestly gave the world the impression that at the base of all measurement, the starting point for all that was precise, were the finished metal tools and instruments made by Joseph Whitworth. The truth is, Maudslay made the first great machines, and then Whitworth made the tools and the instruments and took the measurements that made it possible to make the great machines that followed. The perfectly flat plane was one of them, arguably the essential one.

Two later inventions above all else define Whitworth's legacy. The standardized screw on the one hand, and the measuring machine on the other. Both these creations were linked together, literally and mechanically, and both were connected to the sudden new enthusiasm (around the world, not just in Britain and America) for the newly named science of metrology, the study of accurate measurement. In the years following Whitworth, immense amounts of treasure would be expended all around the world, and still are being spent today, in the pursuit of this new calling, of making officially sure that the measurements of everything around us are all accurate to something, measured to a standard agreed upon by all.

Whitworth's own measuring device was phenomenal for its time, a small thing of the greatest elegance and beauty. It is the sort of beauteous mechanical creation that even a nonmechanical person would wish to own, to look at fondly, occasionally to touch. It can be seen in the portrait of its maker that hangs in the Whitworth Art Gallery in Manchester. He stands in a formal dress coat, with an expression that somehow combines solemnity, pride, and slight surprise. The fingers of his left hand are brushing the brass adjusting wheel, as if to display it, modestly. Beneath, the painter has captured the gleam of obsidian-smooth iron, the instrument's heavy base; other brass wheels glint yellow in the gaslight.

The basic principle of his device is disarmingly simple. Most earlier measuring machines used lines, such as are found on a ruler or a straightedge—one compares the length of something by holding it next to the rule and seeing where, according to the lines, it begins and ends. But this technique requires the use of sight to make judgments, and it raises questions. By how much is the end of the item to the left or right of the line? How thick is the line itself? How powerful is the magnifying glass needed to answer these questions? And even if a vernier scale is brought to bear on the problem—Pierre Vernier made this scale in the seventeenth century to allow one to peer between the lines of the main scale, as it were, and to make ever-more-exact decisions—the answer is still subjective, requiring good eyesight and fine judgment.

Whitworth thought line measurement fraught with problems, as clumsy and liable to error. Instead, he fa-

Standardized screws of differing pitches and thread types, used for fastening, for measuring, and for the advancing and retarding of cutting heads of machine tools.

vored what is called end measurement, which relied not on sight, but on the simple feel of the tightening of the measuring instrument against the two flat end surfaces of the item to be measured. The device he created basically employs two plane steel plates that can be moved toward and away from each other by the turning of a long brass screw. Place an item between these surfaces and tighten the planes until they hold the object securely. Then slowly move the planes away from each other until—the crucial moment!—the item is loose enough to fall under gravity. The distance between wherever the planes are then sited is the dimension of the item.

Measuring that dimension then depends on the screw, on the wheel that is used to turn it, and on the application of simple arithmetic. Consider a screw that has twenty threads to the inch and is moved by a wheel that has, say, five hundred divisions marked around its circumference. Turn the wheel once completely around and the screw, and the plane plate attached to it, advances by 1/20 of an inch. Turn the wheel through one of the wheel divisions only, and the screw advances by 1/500 of 1/20—that is, 1/10,000 of an inch.

Such was the principle. Whitworth, using his superb mechanical skills, created in 1859 a micrometer that followed this idea but that allowed for one complete turn of the micrometer wheel to advance the screw not by 1/20 of an inch, but by 1/4,000 of an inch, a truly tiny amount. Whitworth then incised 250 divisions on the turning wheel's circumference, which meant that the operator of the machine, by turning the wheel by just one division, could advance or retard the screw and its attached plane plate by 1/250 of 1/4,000 of an inch. In other words, by 1/1,000,000 of an inch. And provided the ends of the item being measured are as plane as the plates on the micrometer, opening the gap by that 1/1,000,000 of an inch would make the difference between the item *being held firmly, or falling, under the influence of gravity.* Thus did Whitworth describe the method, some years later, in a paper entitled simply "Iron," and published in New York to a fascinated engineering readership.

The revelation, and the beautiful little machine that

could perform the task so sweetly, astounded the engineering world. Less than eighty years before this, John Wilkinson had given birth to the concept of precision with a machine that could bore a hole to a tolerance of one-tenth of an inch. Now metal pieces could be made and measured to a tolerance of one-millionth of an inch. The rate of change was quite incredible. The possibilities, even if their specifics went then unrecognized, seemed suddenly to be without limit.

All this work was performed in England, most of it in Manchester. Once American machine toolmakers had absorbed all Whitworth's ideas and principles and standards, it seemed probable—and Whitworth, who had been on a fact-finding mission to New York in 1853, was only too well aware of this—that the engineers of the United States would eventually sweep into pole position and propel their country into world leadership. "The labouring classes [in America] are comparatively few in number," Whitworth reported on his arrival back home, "but this is counterbalanced by, and indeed, may be regarded as one of the chief causes of, the eagerness with which they call in the aid of machinery in almost every department of industry. Wherever it can be introduced as a substitute for manual labour, it is universally and willingly resorted to . . . It is this condition of the labour market, and this eager resort to machinery wherever it can be applied, to which, under the guidance of superior education and intelligence, the remarkable prosperity of the United States is mainly due."

And there were the screws—not just the screws that advanced or retarded measuring instruments or microscopes or telescopes, or that elevated naval cannon, but also the screws that held together the parts of all the manufactured goods then made.

Until Whitworth, each screw and nut and bolt was unique to itself, and the chance that any one-tenth-inch screw, say, might fit any randomly chosen one-tenth-inch nut was slender at best. Whitworth championed the idea of standardizing all screws: the threads of all should have the same angle (fifty-five degrees), and a pitch that should likewise be in a fixed relationship to the radius of the screw and the depth of the thread. It took some long while for the individual makers of screws to fall into line, but by midcentury, the standard had been accepted throughout Britain and her empire, and the screw-measuring notation BSW, for "British Standard Whitworth," memorializes him still, as it remains a crucial standard in engineering workshops from Carlisle to Calcutta.

In later years, Whitworth turned his attention somewhat away from the metallic delicacies of high precision and more to the brutish world of weaponry, even though he was vexed that the hexagonally barreled Whitworth rifle that Queen Victoria had fired on that summer Monday in Wimbledon was never accepted for use by the British Army; its .45-caliber size was initially thought too small. He derived some pleasure, though, from hearing that the weapon, branded the Whitworth Sharpshooter in the United States, was much favored by Confederate troops during the American Civil War. (The Union army found his high-velocity guns ideal, but too costly.) His gun was most famously employed with lethal effect at the 1864 Battle of Spottsylvania. The Union general John Sedgwick, seeing the rebel troops in the far-off distance, famously rode in front of his men and loudly declared that "they couldn't kill an elephant at this distance." A single shot from a Whitworth gun then promptly rang out and the bullet hit him square in the head, killing him instantly.

Whitworth may have found his excursion into the military world distasteful, but it proved highly profitable. He designed armor plating and exploding artillery shells and came up with a variety of a ductile steel alloy that he deemed wholly suitable for manufacturing guns—and Whitworth steel, as it was called, became popular among weapons foundries in the United States. In his final years, now with a slew of fine houses at his

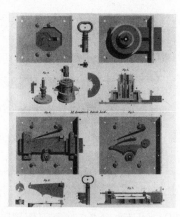

Joseph Bramah's "challenge lock" remained unpicked for sixty-one years after first being displayed in a window in London's Piccadilly. An American named Alfred Hobbs eventually beat the challenge, after fifty-one hours of delicate work, allowing the Bramah lock company to declare its invention essentially burglar-proof.

disposal, and schemes for scholarships and endowments that would keep his name and his legacy familiar today, he designed a billiard table for use in his mansion outside Manchester. It was made of solid iron, and though history does not offer details as to Joseph Whitworth's competence or otherwise at the game, what is recalled is that the surface of the table was renowned for its unique flatness; it was perfectly true. When anyone today bleats about the need for a "level playing field," it is worth remembering that Joseph Whitworth was in all probability the first engineer to give us one.

———

In the closing weeks of the Great Exhibition at the Crystal Palace, in the hall reserved for displays from the United States, an unexpected new exhibit was placed on view: on the floor of a secure glass case was a black velvet cloth, and laid on it, arranged in neat rows, were two hundred newly minted solid gold one-guinea coins. Their unanticipated appearance tells one final story of midcentury precision engineering, one related to the solving of a problem that had been created nearly sixty years before.

A man had managed to pick Joseph Bramah's lock, the very lock that had sat patiently in the front window of the firm's showroom at 124 Piccadilly since 1790. He was a fellow exhibitor at the Great Exhibition, he was a locksmith, he was a competitor, and he was an American. He had come across the Atlantic with the specific intention of picking every unpickable lock that British engineers could place before him.

His name was Alfred C. Hobbs, and he was born in Boston in 1812, of English parents. Maybe that had something to do with his burning passion to demonstrate that American locks were vastly superior to their British-made counterparts.

Upon his arrival at the Great Exhibition, he took up

his position at Stall Number 298, at the eastern end of the main hall, as representative of the New York firm of Day and Newell, makers of the so-called parautoptic permutating lock, which Hobbs was convinced would remain unpickable for all eternity.

Not so with the Bramah lock. Once Hobbs had set out his stall in the Crystal Palace, he wrote a formal letter to the Bramah company, requesting an appointment in Piccadilly "in relation to the offer you make on the sign in the window for picking your lock." Joseph Bramah himself had died forty years before, presumably smugly content that his lock challenge had never been met. It was his sons who now ran the firm, and they received—with some trepidation, as Hobbs's reputation preceded him—the fateful letter. They had no choice but to agree to meet, and a committee of experts was promptly set up to ensure that any attempt on the lock, as precise a mechanism as eighteenth-century England could produce, would be made fairly, and without totally destroying the lock's internal mechanisms.

And Hobbs picked it. It took him fifty-one hours, spread over sixteen days, to raise the lock's hasp and declare it open, and thus successfully broken. He used a variety of tiny and specially contrived instruments to work on the lock's innards—one of them a tiny micrometer screw he was able to attach to the wooden base on which old Joseph Bramah had first mounted the challenge lock. (Had it been mounted on an impenetrable iron base, this instrument could not have worked. It screwed into the wood, thus freeing Hobbs to use both his hands to work inside the two-inch-long lock barrel, while his instrument kept various of the eighteen tiny sliders inside the lock depressed.) He also used magnifying lenses, with brilliant lights whose minuscule beams were reflected inside the lock by means of special mirrors. He used minute brass measuring scales to see how far depressed was each slider. He used tiny hooks to pull back any slider that had been depressed too far. He had laid out beside him what resembled the contents of a surgeon's instrument tray, minus scalpels, for the sole purpose of breaking the Bramah lock and, by doing so, asserting the superiority of American precision.

Bramah paid up, but they grumbled as they did so that what the American had done, with his trunkful of instruments and his fifty-one hours of work, was simply *not cricket.* He hadn't abided by the implied rules of engagement. He had brought to bear on the sorry lock more time and energy than any self-respecting burglar would ever spend.

The team of arbitrators agreed. They pointed out the unfairness of Hobbs's approach, and concluded, ringingly—though well aware that the two hundred guineas had most sportingly been handed over—that "Hobbs has done nothing calculated in the last degree to affect the reputation of Messrs Bramah's lock; but his exertions have, on the contrary, greatly confirmed the opinion that, for all practical purposes, it is impregnable."

The two hundred guineas then glowed impertinently under the lights of the Crystal Palace for many weeks to come, as Alfred Hobbs basked in his victory by insisting they remain *in situ* as testament to his triumph. It was short-lived triumph, and the consequences indicate the eventual outcome. As the arbitrators suggested, the breaking of the Bramah lock did the firm no harm at all: customers lined up to buy a lock that had taken an expert sixteen days to pick. The firm still exists in London today, and sells its locks worldwide, all of them based upon the original design of Joseph Bramah of 1797.

Meanwhile, the firm of Day and Newell of New York went out of business soon after the Great Exhibition. Its parautoptic permutating lock had been successfully picked soon thereafter, and easily, with the use, it was said, of only one wooden stick. And the man who picked it was the scion of a new firm of precise locksmiths, and founder of the firm that is now part of the biggest lock maker in the world, Linus Yale.

Chapter 5

The Irresistible Lure
of the Highway

So profound was the effect of the Model T Ford on
America, so much did it change the nature of the
nation . . . its art, its music, its social structure . . . its
health and wealth and arrogant insularity, that Henry
Ford who was responsible for it all must be seen as the
most effective revolutionary . . .

—L. J. K. SETRIGHT, *Drive On!* (2003)

I was closing the trunk of a borrowed Rolls-Royce Silver Seraph one midwinter's day in early 1998 when
I felt a sudden sharp sensation in my right index finger.

I looked down: a drop of blood was bubbling up from a small nick—insignificant in itself, probably not even worth a Band-Aid. Yet, that some part of a brand-new Rolls-Royce motorcar had been sharp enough to inflict a cut—that did seem worthy of note, not least because the Silver Seraph had been deliberately designed to reinforce, maybe even to reestablish, the idea that in 1998, and despite all the competition, Rolls-Royce Motors still made the very best car in the world.

My co-driver and I checked, running our hands gently across the mirror-smooth surfaces of the rear of the car. There was no doubt this was a beautiful machine: deep blue, with thick wool rugs lining the floor of the trunk, special containers for umbrellas, all the chromium parts thick and solid and highly polished, the lights large and sturdily recessed, even the license-plate holder robust and weatherproof, as if built for a warship.

Except, running my hand along the lower limb of the license-plate holder, I found two tiny screws, and one of them, the right-hand one, seemed to be tilted at an angle such that its razor-sharp steel edge protruded by perhaps a fraction of a millimeter above the mirror-flat surface of the chromium. I ran my thumb across it. There was no doubt; this was the culprit: a simple screw that some apprentice had tried to turn into a hole, that had been ever so imprecisely bored at an angle, fractionally out of true.

For a machine so self-consciously promoted as the motoring world's finest example of precision engineering, and at a price of eye-watering unaffordability to most, this seemed scarcely credible, an unforgivable error, a black mark. My unease was confirmed a few weeks later when a motoring reviewer for one of the London newspapers described how he had taken a Seraph out for a test drive and, after parking it, found not only that could he not release the parking brake, but that the brake handle came off in his hand, its connecting cables having snapped off cleanly somewhere in the bowels of the machine. Clearly someone in the factory was not paying proper attention.

It therefore came as little surprise to me, though of great and shocked dismay to most everyone elsewhere in Britain, that within months, and quite coincidentally, the long-revered institution that had been Rolls-Royce Motors became effectively defunct, and was sold off to the German company Volkswagen.

The company that the world still knows by its hyphenated name, Rolls-Royce (though financial crises and corporate shenanigans of one kind or another have caused there to be all too many versions of the title) was famously founded in Manchester in May 1904. One year previously, in June 1903, and with much-less-remembered ceremony in Detroit, Michigan, the Ford Motor Company had been officially incorporated. Both companies were founded by dedicated, obsessive, oily-handed engineers, both men christened Henry, both born in modest circumstances and in the year 1863.

Once their respective ambitions had jelled, these two men turned out to have markedly different goals. Henry Royce was quite simply committed to building for the discerning few the finest motorcars in the world, no matter the difficulty and with no concern for cost. Henry Ford, on the other hand, was determined to make the world of personal motor transport available to the greatest number of people imaginable, at as low a cost as manufacturing would allow. To achieve their separate ambitions, Henry Royce would assemble a team of craftsmen to build his cars by hand, while Henry Ford would create his cars in immense numbers by, in due course, employing machines to help construct them.

Yet, for both men and both endeavors, extreme mechanical precision was the key—precision, wielded either with the methodical tenderness of an engineer who believed himself an artist, or with the ruthless determination of an engineer who believed himself a revolutionary. A comparison of the two companies will illustrate the manner in which precision, by now, in the early years of the twentieth century, a fully established and essential component of civilized existence, was applied in two very different manners, and with two very different eventual consequences.

It is unlikely now that I will ever have the wherewithal to own a Rolls-Royce, confirming a condition that has been true for all my days so far. Still, I have long admired the machine. Back at university, I was part of a small group that owned a 1933 model, the classic 20/25, that had been hastily and unattractively converted into a hearse. It drove easily and generally ran well, though its fuel consumption was unpredictable, unstated, and, for university students, profoundly unaffordable. We seldom took it out for more than a casual spin. A friend had a harpsichord, which he mounted in the rear, and he would play it as we drove, entertaining passersby. On the one occasion when the car, on a trip into the Cotswolds, actually did break down (or when it "failed to proceed," as Rolls-Royce then preferred one to say), the engineers who arrived to make repairs brought a set of black felt coverings to try to mask the car's identity and save the company embarrassment. This was a largely pointless exercise, and fooled nobody: people would see the felt pads on the "RR" hubcaps and would spy the black tea cosy–like arrangement that more or less covered the Spirit of Ecstasy hood ornament and the Grecian-style radiator on top of which she stood, and would recognize instantly what marque of car was in trouble.

My fondness for the motorcar took fuller flight some years later, in early 1984, when I was given an assignment for a London newspaper to write a number of essays about mainland Europe—about which, an editor cynically remarked, the average Briton knew little, and wished to know less. The essays were each to be reports of my chosen journeys among a variety of cities and made in a variety of ways. So I took a boat from Stockholm to Helsinki; I walked from Cádiz to Gibraltar; I took a train from Victoria Station in London to the Hotel Victoria in Brig, on the Swiss-Italian frontier; and I was to drive a car—this was intended to be the assignment's cover story—from the westernmost point of Europe to its most easterly, from the headlands of Atlantic Galicia to the then-Soviet city of Astrakhan, where the Volga meets the Caspian Sea.

I left this epic car journey until last, once I was done with the sailing and walking and train riding. Initially, I had been planning to take the old family Volvo on what would be a many-thousand-mile odyssey—except that, toward the end of what must have been a rather too fortifying lunch in central London, I had wondered out loud to Patrick, the photographer who was making the expedition with me, why not take a Rolls-Royce? It might make quite a stir in the Soviet Union, after all.

It was all too easy. A swift call to the company's PR department, and all was fixed in no more than half an hour: a Silver Spirit in Ocean Blue would be coming off the production line the following morning—a canceled order—and if I could trouble myself to take a train up to the factory in Crewe, the car could be mine for the next two months. "Bring it back in one piece is all we ask," said the PR man the next morning as he gave me the keys. We shook hands, and Patrick the photographer and I drove off.

The adventures that befell us on what turned out to be the most epic of journeys do not belong in these pages. The precision of the car's inner workings and the fastidiousness of those who had prepared it for the voyage were such that the eventual ten thousand miles of driving were accomplished impeccably, in perfect and quiet comfort, at high speed where necessary—on occasion, in Bavaria, at as much as 140 miles per hour, no small speed for a three-ton car—and without even the most trivial of mechanical incidents. The only visit to a mechanic came when I met the dealer in Vienna (in those days, Rolls-Royce's most easterly outpost) to have the engine's timing slightly altered to take account of the low-quality fuel we would likely encounter beyond the Iron Curtain. "Though, quite candidly," said the dealer, patting the warm cylinder head, "this engine could run happily on peanut butter, it's so accommodating."

The essays duly ran in the newspaper, with, as expected, the Rolls-Royce trip chosen as the cover article, mainly because of the emblematic accompanying photograph. This showed me hamming it up outside the city gates in Kiev. In it, I am sitting on the hood of the mighty blue car, which has been freshly polished and is showroom-gleaming and therefore the embodiment of rich, vulgar capitalism. I am pointing at something or somewhere in the middle distance. What made the image coverworthy was that Patrick had set our Rolls-Royce directly in front of a huge agitprop painting of Comrade Lenin, who was standing, chest out and legs apart in manly fashion, pointing his index finger at the same uplifted angle as my arm, and also into the same middle distance, to a mythical destination that was presumably, for the people of Kiev, the brave and glorious future of the USSR. The contrast could not have been better chosen, the irony inescapable. The publication issue sold well in London; it was banned, I think, in Kiev. The brief local success of the piece spawned a decade of unanticipated gratitude and generosity from the public relations staff at Rolls-Royce Motors, worldwide.

My very next assignment for the newspaper was to write an essay on the gangs of East Los Angeles, as the 1984 Olympic Games were about to open and the local authorities were said to be fretful. Accordingly, I flew to California with another photographer and, upon checking in to the Ambassador Hotel on Wilshire Boulevard, was more than a little surprised to be handed a small brown envelope with a letter inside from Rolls-Royce of Beverly Hills, and a set of keys. "Enjoy your stay," the letter read. "This is on us."

"This" was a brand-new behemoth of a car, a black-and-white Rolls-Royce Camargue, the most expensive production car in the world at the time, and one of the least attractive. It was a two-door monster of a machine designed by an Italian on what was evidently a very bad day. It was slow and cumbersome and heavy and a classic example of automotive mutton dressed as lamb, and as such, it attracted much unwanted attention. I was waiting at a red light one hot afternoon when a pair of young women in a convertible drew up alongside. "That a Rolls-Royce?" the driver asked. "Yes," I replied. She laughed. "Ugliest fucking car I've ever seen."

The story of the Camargue amply illustrates a difference between precision and accuracy. For while the engineers had lovingly made yet another model of a car that enjoyed great precision in every aspect of its manufacture, those who had commissioned and designed and marketed and sold it had no feel for the accuracy of their decisions. As a result, the Camargue was a serious commercial flop, the Edsel of Crewe. The company, just then starting the slow decline that would end with the nick on my finger and the snapping of the brake cable and the transfer to German ownership a decade or so later, sold just a few more than five hundred Camargues over the ten years the model was in production. In 1985, the year after I had my two-week loan of it (an unsold and unsellable model from the Beverly Hills back lot, I came to realize), the company put it out of its misery and shut Camargue production down for good.

Had there been more justice in the world, the company would have been named Royce-Rolls, as Henry Royce was the man who made the cars, while Charles Rolls simply (and flamboyantly) sold them. Yet, with the name known for years as one of the most familiar brands of all time—only Coca-Cola is said to have been better known—the notion of altering it by even the most infinitesimal degree has long been considered a sacrilege. The hyphen, for instance, is sacrosanct. The diminutive use of "Rolls" is said to have been regarded as inexpressibly vulgar. The men on the shop floor, if pressed to speak about their creations in familiar terms, called them "Royces."

It was all to the good that Henry Royce was born near Peterborough,⋆ where, soon after his birth in 1863, the Great Northern Railway happened to have built a locomotive repair and maintenance workshop. For although his childhood was both impoverished (in his youth he was obliged to work variously as a bird scarer, newspaper seller, and telegram delivery boy) and harsh (he was just nine when his father died, and in the poorhouse), he had an aunt who with great prescience believed that to learn the trade of engine building would set the boy on course for life. So she paid for three years of apprenticeship for young Henry in the Great Northern Railway workshop, a place that would soon go on to build and repair some of the finest and swiftest ever of Britain's steam locomotives. And just as she had hoped, her decision to pay her nephew's fees set him on course for making engines himself. Though, to be sure, his would be motorcar engines that would come to enjoy great repute,

---

⋆ Might there have been something in the water? The Cambridgeshire village of Royce's birth, Alwalton, was also where Frank Perkins, inventor of a much-revered brand of diesel engines, was born twenty-six years later. Yet only Henry Royce won the memorial plaque in the local church.

and they would be of much greater mechanical delicacy than the ironbound, coal-gorging monsters on which he had trained and numbers of which he had helped bring out from the Peterborough railway sheds.

It would, in fact, be more than twenty years before Henry Royce came to make engines himself, and the motorcars to put them in. His first venture involved electricity, and was out of a workshop on Cooke Street in Manchester that manufactured and sold such new-fangled devices as light switches, fuses, doorbells, and dynamos. He soon became moderately prosperous, married, bought a fair-size house in the suburbs, and embarked on the devotion of such spare time as he had to gardening and raising roses and fruit trees, a passion for the rest of his life.

Mechanical rather than electrical engineering was his true passion, though, and within a decade he took steps to incorporate the two, setting up a company named Royce Limited that produced a range of large-scale industrial electric cranes. The firm won a following and a fine reputation: its cranes were known to be both very well made and built with patented Royce-designed safety features that minimized the number of lethal accidents that were then plaguing the new world of Victorian highish-rise construction. Over the years his company flourished, even selling electric cranes to the

Imperial Japanese Navy, and having one exactly copied by unscrupulous Japanese engineers, right down to its ROYCE LIMITED nameplate.

Around the turn of the century, a number of German and American companies suddenly entered the crane market, undercut the Royce prices, and nearly brought the company to its knees. Royce, in an early display of a case-hardened determination to make machinery of the highest quality whatever the pressure, insisted he would neither cut his costs nor trim his standards—and in time, the young company survived, stabilized, and gained a reputation for high-quality engineering, for precision products made beyond consideration of price.

If life were fair, the motorcar Henry Royce created in 1904 would be named the Royce-Rolls, as the latter, Charles Rolls, was little more than a salesman and promoter. In the machine shop, the engineers defiantly called their creations "Royces."

Henry Royce was by now himself settled, stable, domesticated, and with money in the bank. His personal interest turned to automobiles. He was able to indulge himself—first, by buying, in 1902, a De Dion quadricycle, essentially two bicycles bolted side by side, with a small internal combustion engine suspended between them. France at the time had a near monopoly on the stripling car-making business, with firms such as De Dion-Bouton, Delahaye, Decauville, Hotchkiss et Cie, Panhard, and Lorraine-Dietrich producing small numbers of vehicles for a growing number of enthusiasts. The vocabulary reflects still the Gallic origins: words such as *garage, chauffeur, sedan, coupe,* and, indeed, *automobile* serving as reminders.

Henry Royce thought at first that the French cars were good-looking and admirably well made, and by craftsmen—and were far better finished than the rather cruder American cars that were also starting to appear on European roads. He soon began to take a more serious interest, and in early 1903 he purchased his first true motorcar, a secondhand ten-horsepower two-cylinder Decauville, which arrived in Manchester on a train and had to be pushed by Royce's workers from the railway station to his Cooke Street workshop.

The "10 Horse Standard" was a state-of-the-art car of 1903. A dealer in London keenly advertised its recent achievements: "Edinburgh to London Without Stopping! Average speed of 20 miles an hour for the whole distance with a full load!" "51 Miles an Hour at Welbeck!" "75 Miles an hour at Deauville!" He claimed the car could travel up to thirty-five miles an hour in average conditions, could carry four people in comfort—an extendable cover for the tonneau passengers provided them shelter from the rain, though there was no shelter for the driver, nor any windshield—and petrol was just a shilling a gallon, "always available" at the dealer's chambers.

It was within only a few weeks of buying the car that Royce made a fateful decision. He was enjoying his machine, and he drove it almost daily, but while the design was acceptably chic, the mechanisms inside, in his view, turned out to be sorely wanting. The car was noisy. Its acceleration was poor. It overheated easily. It was not in the least reliable.

He promptly announced to his team that he would strip the car down to its bare essentials and redesign it from the wheel treads up, creating in the process an entirely new kind of car that would be, in every respect, mechanically perfect and utterly reliable. He would do the initial work in his own time, and if what he then created seemed at least the beginnings of an ideal, he would set Royce Ltd. to manufacturing entirely new cars based on his redesign of the French machine, and he would call it a Royce. The Royce 10 horsepower. The Royce Ten.

Painstakingly, and almost entirely by the employment of a delicate hand and a steady eye, the new car took shape. Like the Decauville, the Royce had two cylinders, each with a bore of 95 mm and a stroke of 127 mm. The fuel inlet would be at the cylinders' top, the exhaust valves on their side. There would be a water-cooled jacket at the front of the engine, ensuring that the machinery never overheated. Royce designed and hand-made a new kind of carburetor; he made a new wooden-cased trembler coil, with hand-finished points of pure platinum, which never seemed to need either adjustment or cleansing, and from which came the ceaseless rain of high-voltage sparks to ignite the fuel. Usually it was the coil that gave the greatest amount of trouble in a 1904 car; Henry Royce's gave, at least in this department, none. Moreover, Royce also fashioned a highly accurate distributor, which made certain the cylinders were ignited at exactly the moment they received the jolt of the petrol-and-air mixture that keeps an internal combustion engine running.

He introduced a driveshaft instead of a chain drive. He saw to it that every gearwheel fit perfectly and was lubricated generously. He perfected the car's suspension, always aware that people would be riding in this vehicle, and had to be kept both comfortable and safe. He fashioned cylinder-head gaskets from the leather of his apron. He designed tapering bolts that would replace the rivets of the French design. He made an enormous, overgenerous, multibaffled silencer for the exhaust system, so determined was he to cut the exhaust roar to no more than a dull murmur. His gearbox had three forward speeds, and the clutch was lined with leather. He replaced the worm gears on the steering system and the shoes on the braking system—and by then testing countless times, by analyzing every breakdown, he made certain that his Royce Ten would be a more than acceptably reliable alternative to the now-picked-over carcass of the sacrificed Decauville, albeit vastly costlier by being so. "No wear, frettage or indication of malfunctioning was too trivial for him to notice," said one of his later and more famed engineers, Sir Stanley Hooker, "and to make efforts to correct."

The first Royce Ten emerged from the Cooke Street workshop on March 31, 1904. In short order, two more machines, each one better and more finely constructed than its predecessor, were rolled out into the street. Then a new board member of the Royce company, a man named Henry Edmunds, photographed one of the gleaming new cars and sent the picture down to London to a friend—the friend being the Honorable Charles Rolls, a leisured near-beer aristocrat, daredevil, showman, and car enthusiast (and member of the Self-Propelled Traffic Association) who at the time was trying to sell Peugeot and Panhard cars to rich customers in the quietly exclusive streets of Mayfair, Knightsbridge, and Belgravia.

On receipt of the small black-and-white snapshot, Charles Rolls was mesmerized, instantly electrified. He realized, from Henry Edmunds's description and from this one picture, that at long last a British car of merit equal or superior to those from Continental manufacturers could now be had for the asking. He wrote to Royce, initially asking, then demanding, then begging that this most extraordinary mechanic come down to London to meet him. He wrote letter after letter. Each, however, was rebuffed.

I like to imagine the scene in Cooke Street in late April of that year. There was yet another letter sitting on Henry Royce's desk, but one that, yet again, the engineer had had no time to answer. The letter had come from London; now it was in Manchester, and Henry Royce knew it would be yet another plea from this metropolitan swell, this Old Etonian and Cambridge graduate, pleading for Henry Royce to go down to London for a meeting.

But Royce was not planning to budge. He was far too busy, and the work he was performing in his cramped little mechanical shop was consuming his every waking moment.

All of the previous early-spring week, I like to suppose, he had been working on a near-impossible self-imposed task: he had been trying to machine a forged-steel crankshaft into such perfect balance that, once set spinning, it would never stop, as no one side of the shaft would be heavier than another, which would have tended to slow down the spinning. On the day the letter from Mr. Rolls arrived, he was fiddling with a micrometer, trying to measure the tolerances of the oddly shaped shaft, polishing and filing its extended segments until his gauges showed they were no more than a hundred-thousandth of an inch different from one another, were essentially identical, were as perfectly balanced as it was possible to make them.

Henry Royce was fully enraptured with building his motorcars. They each, he told his workers, would in the end, after trials and testings and endurance exercises and rebuildings, be cars like no other. The components of each would be so lovingly sculpted, machined with such unyielding accuracy, that the resulting cars would be eternally reliable, whisper-quiet, intensely powerful, and, to devoted engineers if not necessarily to the general public, things of consummate mechanical beauty.

So now, with the crankshaft made and tested (and indeed, made with such perfection that, once spun up to speed by hand, it manifestly did not want to stop), the latest version, the third, of his Royce Ten passenger-carrying, fully motorized, and entirely English-built conveyances was ready to be tested, to be driven. The completed engine was bolted onto the chassis. The drive train—also a thing assembled by hand, its component parts polished with chamois leather until they gleamed and flashed in the afternoon sun—was connected. Wheels with pneumatic tires were bolted onto the axles. Fuel was carefully poured into the tank.

Royce then inserted the nickel-steel hand crank into the slot beneath the cooling radiator, a radiator whose Grecian-looking temple top gave the brand-new automobile an unusual appearance of dignified nobility. He turned the crank once, twice, three times.

At first, nothing. Royce adjusted a lever, turned a knurled brass wheel, opened a valve a little more. And then, with a series of low grunts and an initially alarming burst of dark smoke from the engine, exhaust that made the workers step back in alarm, the motor caught, fired, and then promptly settled down to a low thunder of rotation.

The engine was so very quiet. It didn't make the raucous and tinny din like the Decauville. No, this one was something else. The exhaust burbled gently. The tappets clicked near-silently. The camshaft lifted and lowered the valves with the silky sound of well-oiled metal. Once the bonnet—a new word, invented for the motorcar's hood only the year before—was closed and secured around the shuddering engine, it fell truly silent, and only its heat and the hand-sensed feel of its vibration assured the awestruck engineers that it was still firing—"firing on all cylinders," as the phrase would soon have it.

The test driver then clambered aboard; adjusted the choke, his cap, and the magnetos; and set his goggles over his eyes. Someone opened the double wooden doors of the works and glanced up and down Cooke Street to make sure it was clear of passing horses and pedestrians. The driver eased the transmission into first gear, released the brake, grasped the steering wheel, released the clutch—and Henry Royce's third handmade motorcar near-inaudibly slipped out into the street and glided off for the low hills on the horizon, starting its own first real-world expedition.

It was then that Henry Royce opened the envelope.

It was indeed a letter from Charles Rolls, but on this occasion, there was no plea that Royce go south to London. On the contrary: if it was convenient, Rolls himself would come up to Cooke Street, and would come to see if it might be possible to manufacture and sell the best car in the world. Might Royce be amenable? Both parties, the letter said, revered the notion of making a superior motorized passenger conveyance that would be built on the principles of absolute precision, no matter the cost. Might Henry Royce think of the writer and himself one day going into business together, and perhaps calling the new company by some arrangement of their commingled names?

Two hours of test driving later, the little black car swept back onto Cooke Street, its still-ghost-whispering engine now bathed in warm oil, its driver astonished and delighted by its performance. All reports from the journey were peerless. The car had exceeded all expectations. And so, that evening, emboldened by the evident success of his new creation, Henry Royce wrote back to Charles Rolls. By all means come up to Manchester, he said, and we will meet on May 4, two weeks from today. Perhaps, after all, we may be able to do business together.

There is a brass plaque outside the entrance to the Midland Hotel on Peter Street in Manchester that memorializes the first time Charles Rolls and Henry Royce formally met, as scheduled, on May 4, 1904. All that Royce was hoping for from the meeting was the funding to allow him to continue making motorcars of ever-more-demanding exactitude. What Rolls wanted, as he told Henry Edmunds on the railway journey up from London that morning, was for his name to be associated with some great creation such that he might in time become a household word, "just as much as 'Broadwood' or 'Steinway' in connection with pianos," wrote Edmunds, "or 'Chubbs' in connection with safes."

The sight of a brand-new Royce Ten, and of Henry Royce's evident quiet pride in his having made it, did the trick—as did a short, smooth, faultless ride through the streets of Manchester in a car that manifestly did not frighten the horses. (The terrific noise of most others did.) Rolls returned to London that night by train, and evidently supped well in the dining car, because he was out and about at midnight proclaiming to all in Belgravia who would listen, "I have found the greatest engineer in the world! The greatest engineer in the world!"

The lawyers got to work the next day, a formal deal was struck two days before Christmas, and the partnership formally came into being. There had been no difficulty in securing the order of their names for the new firm. Henry Royce happily let pride in the quality of his machines supplant any fuss about nomenclature, so it was readily agreed that it would be "Rolls" first and "Royce" after, conjoined by a hyphen: Rolls-Royce. Rolls-Royce Limited.

At a dinner in 1905 that followed the first-place win of a Royce car in a race staged on the Isle of Man—and yes, automobile production began with almost indecent haste just as soon as the company had been formed; the winning of competitions provided a perfect publicity device—Charles Rolls told of his first meeting with Royce. He had been trying to peddle French-made cars to the London beau monde, but then:

> I could distinctly notice a growing desire on the part of my clients to purchase English-made cars; yet I was disinclined to embark in a factory and manufacture myself, firstly on account of my own incompetence and inexperience in such matters, and secondly on account of the enormous risks involved, and at the same time I could not come across

any English-made car I really liked . . . eventually, however, I was fortunate enough to make the acquaintance of Mr. Royce, and in him I found the man I had been looking for for years.

The very earliest cars made at Cooke Street were known not so much for style or speed or strut or panache as for their quiet and their reliability. A decade after the first handmade Tens nosed out of the factory came the stories of endurance. A farmer in eastern Scotland, for example, had run his Ten for over one hundred thousand Highland miles with not a single breakdown, and his car hadn't been all that expensive: Royce charged £395 for a Ten, at a time when a sixty-horsepower Mercedes cost £2,500 and a six-cylinder Napier a little more than £1,000 (£100,000 in modern sterling, $128,000 in 2017 prices).

It was not all beer and skittles, though. There were occasional setbacks. Charles Rolls took one car across to the Isle of Man and decided, imprudently, to coast it down a long hill in neutral. He then, even more imprudently, tried to engage his gears, but forgot to match his engine's speed to that of the coasting car and, in the process, ripped the gearbox apart and stripped the gears down to toothless bare metal. Henry Royce was not best pleased, but gritted his teeth: after all, the Honorable Charles Rolls was known in the company, and not for nothing, as Number One.

It is tempting to suppose that the more modern cars—the mighty monarch- and emperor-carrying Phantoms and all their Silver-prefixed siblings; the Dawns; the Wraiths; the Clouds; the Shadows; the Spirits; the Spurs; the Seraphs; and maybe even the non-Silver-prefixed Corniches and theCamargues—were the company's finer creations. It is tempting to suppose that these cars' legions of refinements and new technologies and layers of opulence and comfort—with many essentially meaningless improvements, such as the use of unblemished hides from cattle kept well away from barbed-wire fences, a suspension system that automatically compensated for the steadily lessening weight of fuel in the rear-mounted tank, Axminster carpets so thick that a lost earring might never be found, dashboard marquetry made for the most elegant of city drawing rooms, and door trim veneers chamfered from ancient trees and matched so they appeared as if in mirrors—exactly reflected Henry Royce's dream for the eternal and the immaculate.

Yet not so. Engineers are a breed hardly known for favoring opulence and vulgarity, and they care little for ankle-deep carpets or butter-smooth leather. They would rather employ their skills to push the boundaries of mechanical possibility, and in terms of making motorcars, that means the use of better materials with a goal of ever-increasing lightness and efficiency, and with the achievement of ever-finer machining tolerances, greater smoothness, more polish, better *fit*.

Up to 1906—and yes, this was still early days in the company's history, but Rolls-Royce was self-evidently moving itself along very fast—every one of the company cars had been based on that original Decauville ten-horsepower car from France. Henry Royce had made all too many versions of this—the Ten, the Twenty, the Heavy Twenty, the six-cylinder Thirty. They were well received by the motoring press, and they sold handsomely, but in engineering terms, they represented to the envelope-pushing technicians something of an intellectual dead end.

What was needed now was an entirely new car, one based solely on Henry Royce's imagination and which owed very little to a now-somewhat-outdated Gallic import. So the company's small band of craftsmen—with their sand-brown grease-stained overalls, their wads of cotton waste, their oil-grimed fingers, their hooded eyes and furrowed brows; with their loupes on lanyards, their slide rules, micrometers, calipers, verniers, and pressure gauges; and with their well-bitten pipes clenched between tobacco-yellowed teeth—stayed late into the 1906 nights, poring over blueprints and log tables, over lists of new alloys and charts that told of the density and flexibility quotient of possible ashwood chassis frames, over screw threads and tappet clearances and potential cylinder diameters . . .

The model that resulted from all this ferment was to be the original Rolls-Royce Silver Ghost, which was first made in 1906 and continued in production until 1925. Nearly eight thousand were built, and most of them are still running today. The car was enormous and, to power it along, had a truly massive six-cylinder side-valve engine, drawing more than seven liters (seven and a half liters from the 1910 models). Everything about the engine was massive, solid, had *heft*. The cylinders were arranged in two cast-iron blocks of three, rounded at the top and finished in brass. There was a single camshaft, exposed tappets; there were copper pipes bringing in the fuel, a twin-jet carburetor with a governor that could be set from a control on the steering wheel; and there were enormous copper tubes to carry the exhaust away to the tailpipe. The crankshaft was polished steel and had seven bearings. Even today, a Ghost engine manages to look both sophisticated and elephantine, as though a marine turbine has been bolted onto a motorcar frame, offering it much more power and endurance than it could ever need.

The car is regarded still today as the nonpareil, the exemplar of all that is right about engineering accomplished to the very highest of standards, and with the highest level of precision. What sets this one model of car apart had more to do with endurance and reliability, quietness and speed, than with excess. "Perfection," begins one of Royce's better-known apothegms, "lies in small things." But perfection is no small thing, and from radiator to tires, carburetor to brakes, the Silver Ghost amply reflected this.

The car was originally called the Rolls-Royce 40/50.

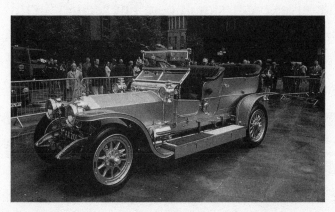

The Rolls-Royce Silver Ghost remains the iconic version of the famous marque, and was the only model that, for a time, was also manufactured in the United States (in Springfield, Massachusetts). Almost eight thousand were hand-built between 1906 and 1925.

The decidedly unromantic notion embodied in this first-chosen name has all to do with regulation, and with that direst enemy of motoring joy, vehicle taxation. Cars in the early part of the twentieth century were taxed according to their horsepower, a calculated number that was decreed by the mandarins of the Inland Revenue in London as being "two-fifths of the square of the engine's cylinder diameter in inches, multiplied by the number of cylinders." This car had six cylinders, each with a diameter, or bore, of about four inches. Four squared is sixteen, six times sixteen is ninety-six, and two-fifths of that number is more or less forty, lending the machine the taxable horsepower figure of forty.

That gives the first number. The second (in this case, the fifty) is the actual horsepower that the carmakers (in some cases boastfully, but usually not) believe or claim their machine is capable of generating. So the two numbers together, the *taxable* horsepower followed by the *actual* horsepower, give us, for this particular car in 1906, the number name "40/50." A more tedious name for a motorcar of such pretension could scarcely be imagined.

Then came a moment of inadvertent marketing genius. After making the eleventh chassis for the new series, the managing director at Cooke Street, the "broad shouldered extrovert and party-giver" Claude "CJ"

Johnson,* ordered that the coachwork of the twelfth (numbered 60551) be painted in silver enamel and all its brightwork made of solid silver, with the intention that the machine be used as a demonstrator. Johnson then named this one particular model the Silver Ghost because of the car's appearance, he said, of "extraordinary stealthiness." The name was hammered onto a plaque, repoussé-style, and mounted on the rear of the car's scuttle.

Matters might have rested there, with merely one car sporting the name, but the influential motoring paper *Autocar* took the view that the entire line could and should have the same name. So, while the factory—and within a year, Cooke Street had been abandoned in favor of a brand-new purpose-built plant in Derby—continued to hand-make 40/50s, both the buying and the admiring public took the Ghost name and formally entered it into motoring history.

---

* Johnson always considered himself "the hyphen" in "Rolls-Royce," that he was essentially the godfather of the Silver Ghost, and insisted that the firm make only one model and make it as perfectly as possible. Considering he was the man who gave the car its name, the founder of the Royal Automobile Club, and, one might argue, the first to make motorcars popular in the United Kingdom, suggests that his importance was far greater than as a mere bridging symbol.

The demonstration car performed wonders. It was taken from the production line on April 13, 1907, driven for eighty miles by the company's chief tester, and then, with all declared shipshape and Bristol fashion, sent on by road to London and into Claude Johnson's custody. He then arranged for it to be subjected to a series of witheringly difficult trials with ever-vigilant men from the Royal Automobile Club observing, looking for any moments of failure. There were, essentially, none—except that there were punctures every few dozen miles, tire failures that were regarded by motorists as little more inconvenient than having to stop for fuel.

On one run over the five hundred miles from London to Glasgow, the car was kept in third or fourth gear only. There were two reasons for this. The first was to test the power of an engine that was being asked to climb great hills, most notably the immense pink-granite lump of Shap Summit, in Westmorland, a notorious grind of a climb back when the A6 was still the main highway to Scotland, then a narrow road to the deep north, indeed. The Ghost glided up it with consummate ease, and then swept down the northern slope at sports car speed. The second reason for keeping the car in one gear was to show Edwardian drivers how easy it was to drive it—a startling number of car buyers had no idea how to change gears, and were terrified at the prospect of having to do so. (Until quite recently, a Rolls-Royce owner's manual would assume the presence of a chauffeur. "In the event of a flat tire, instruct your man to pull over to the side of the road." Almost certainly "your man" would know how to change gears, and tires.)

The Silver Ghost expedition that truly impressed the reading public, however, and that inarguably made Rolls-Royce famous up and down the country, and for all time following, was an endurance test, a trial to see just how far the car could be driven without stopping. It began in June 1907, and was almost casually tacked onto the end of a previously organized romp through the Scottish Highlands, where the car, with Claude Johnson at the wheel, two passengers, and the RAC observer, bumped and ground its way in drenching rains across eight hundred miles of unpeopled scenic majesty. There was a hiccup here: on day one, coming up from Glasgow en route to Perth, the car successfully negotiated the infamous Rest and Be Thankful Pass, but in attempting to round the Devil's Elbow on day two, the tiny brass gasoline tap shook itself shut and promptly starved the engine of fuel, stalling it to a dead stop. A moment's mystification, and then a swift turn of the tap, and all was well again—a foolish embarrassment, but hardly much cause for dismay.

Otherwise, all was faultless, and the car collected an

omnium-gatherum of awards and medals after its five days in Scotland—whereupon Johnson, eager to make as much of a splash in the papers as he could, persuaded the hapless RAC man to stay aboard, and turned south for Glasgow, bound for London yet again. They went, by way of Edinburgh, Newcastle, Darlington, Leeds, Manchester, and Coventry, to the RAC clubhouse on Piccadilly. Then they turned around and headed back north once more, eventually doing so no fewer than twenty-seven times more. The car seemed to love it, simply refused to quit. The RAC inspector and various members of the motoring press stopped by to see the machine slide its way back and forth across England and Scotland like a shuttle on a lanolin-slicked loom, back and forth, back and forth.

Stunts that now seem routine were performed for the first time: a penny piece was balanced on its edge on top of the radiator, the engine was revved to full power, and all professed awe as the coin remained upright, imperturbable and undisturbed. Likewise, a brimful wineglass, together with a freshly made martini lapping its meniscus against a frosted rim, were positioned on the radiator's pediment. The driver was instructed to press the accelerator to the floor and let the full thrashing power of the six-cylinder monster do its worst. In the glassware: not a ripple, not a swish, not a spill. The martini was neither shaken nor stirred by the wrath of the engine, and was afterward said to have tasted fine.

The 40/50's engine was so quiet, said the man from *Autocar*, that it was as though a sewing machine had been hidden beneath the hood. Even though it had the looks of a thumping marine engine, its full-throttle sibilance suggested that within the bowels of the car there lurked a device made for threading slivers of waxed cotton through a chemise of light silk. It most certainly was not the sound of a juggernaut built to power six thousand pounds of automobile and four bearded and burly passengers uphill through a drenching nighttime downpour at eighty miles an hour.

Claude Johnson called a halt to the driving test only on August 8, after forty days of nonstop running, after 14,371 miles had been run without a single involuntary stop—aside from the shut-gasoline-tap stall back up in Scotland, and aside from the halts caused by blowing tires, which had a tendency to fail interminably and inevitably. Servicing of the car had to be accomplished at night, when the drivers were asleep. The only serious work, scheduled before the team left, had been to grind the valves—it was a job that took eight and a half hours, and like most of the procedures that involved Rolls-Royce cars, it was done by hand, slowly, meticulously, and perfectly.

And then the marathon test was over. Now, with the car cooling and creaking and resting in London, Johnson demanded it to be stripped down to its essentials and rebuilt, as new. So every panel and portion of fascia and marquetry piece was removed, RAC men hoisted the enormous engine from the chassis, the transmission linkage was disassembled from the wheels and the gearbox, the brakes were dismantled, and the electrical equipment disconnected. Then a small army of men with micrometers fanned out. Each measuring device had its calipers set to the exact dimensions of the Ghost when it had been delivered on April 13, some 117 days before.

In the engine, the gearbox, and the brakes there was not even the slightest evidence of wear. There was no measurable difference between the engine's condition back in April and now in August; between the state of the car's most crucial components when they were new and now, after they had been hard driven and manifestly well used. Bringing the car back to its original condition required only "the replacing of two front wheel pivot pins, a steering rod tie pin, the ball tip of the steering lever, the magneto driving joint, a fan belt and a petrol strainer. The steering ball joint's sleeve was refitted and the valves were reground."

The RAC report stated unequivocally that had this car been in the possession of a private owner, none of the work would have been either needed or undertaken. As it had now been done for the RAC, however, a bill had to be sent: the total cost of the necessary parts and labor after the Silver Ghost's fifteen thousand miles of arduous travel was a scant twenty-eight pounds, five shillings. A Rolls-Royce, headlined the newspapers, was so indestructibly well made that it might almost be said to be economical to buy, its purchase an investment. There was much fuss in the magazines, with pictures and eyewitness accounts seemingly everywhere.

You could buy a Ghost chassis alone (the frame, wheels, and machinery) for £980, initially. Over the twenty years when Ghosts were manufactured, the price rose to an eventual £1,850 in 1923. A total of 7,876 Silver Ghost chassis were manufactured. So popular were the cars among American buyers that a factory was opened in Springfield, Massachusetts—the city where, one might recall from earlier in this story, mass production began, though of guns, not cars—and in both factories, in Derby and Springfield, the actual method of making the cars was much the same, time-honored, customary. It was a method of making that would be profoundly different from the way Ford motorcars would be made, at almost exactly the same time.

A plan was first chalked out on the factory floor; the iron and ashwood parts of the car's frame were then welded and bolted and riveted over the template, all the pieces propped and supported on stanchions until the moment that the axles were swung down from above and the wheels attached, after which the assembly could stand in one place on its own four wheels, wooden chocks preventing it from moving.

An overhead traveling crane would then bring in the engine, already assembled, for the most part, by hand in a distant part of the same factory. It was a heavy thing, tricky to maneuver, but it would be lowered carefully into position just behind the front wheels, after which the transmission and the gearbox and the universal joint and the propeller shaft and the connection to the rear axles would be made to fit in behind it. The steering gear and linkage would then be hand-assembled and bolted into place on the front wheels and connected by worm gears to the steering wheel, which would be placed behind the engine and to the side of the great gearbox, with its shift lever and its three and, later, four forward gears. The brakes would be shimmied into place, and the levers and linkages and, in time, the slender hydraulic pipes would be connected and sealed and filled with fluid. The batteries would be connected; anacondas of electrical wires would be folded around the engine and along to where the lights and the horn and the various indicators were positioned.

The radiator, that emblematically Greek columnar front end of the car that remains its most recognizable component still today, would have been welded and brazed and polished by a man who had been doing the same thing for all his time with the company. It would then be gently and lovingly and, in truth, reverentially brought to the front of the new machine and bolted into place, polished once again, and connected to the cooling systems, and with the fan to draw air through its silver vanes to keep the engine water from boiling. Lubricants of various kinds and viscosities would be pumped and poured and injected into a variety of locations within the fast-complicating mess of mechanicals, until the moment when fuel would be poured into the tank, the crank would be turned, and the new engine would cough and splutter and start, and then quiet itself to a low murmur. In the early days, all the workers in the factory would stop for a moment to hear its purr and think of it as a newborn, and they the parental team, proud and thrilled.

And then the men from the coach-building companies (usually Park Ward, H. J. Mulliner, J. Gurney Nutting, Barker, or Freestone and Webb) would come take the chassis away and add the carefully sculpted body and the veneers and the carpets and the glass and all those additions that interest the engineers so comparatively little, but that attract the customers far more than the components that actually make the whole confection *work*.

And all that was left were the chalk marks on the factory floor, and in due time, another set of hollow-steel struts would be laid out on top of the template they had made and would be bolted and riveted together as before, and then axles inserted and the assembly lifted onto its wheels and yet more parts would be brought to it and confected into yet another car—and the whole process, slow and painstaking and reverential and shipyard-like, would begin again, and in due course, another Silver Ghost chassis would slide through the doors, eight thousand of them over the subsequent four thousand working days of the eighteen years of that model's production, at a rate of two cars a day. Just two cars a day.

The year after the nonstop running experiment had been successfully conducted, and when all the fuss and bother of its achievement was dying down, Charles Rolls purred his explanation for the car's success. Why, he was asked, did his factory, so fully equipped and manned as it was, not simply produce thousands of cars? Why just two, when it was possible to produce two hundred, or two thousand?

In the first place the class of man who would be quite acceptable in ordinary engineering works would be

*quite unsuitable for us and for our standard of work
. . . To produce the most perfect cars you must have
the most perfect workmen, and having got these
workmen, it is then our aim to educate them so that
each man in these works can do his particular work
better than anyone else in the world . . . We have
always believed that the construction of a motor
car which, while possessing every degree of necessary rigidity and strength, was of less weight than
other similar cars, is largely a metal question. We
consider that the success of the Rolls-Royce and its
extraordinary durability and low cost of upkeep, as
exemplified in the 15,000 miles trial of last year,
is entirely due to scientific design, to the original
research work and close study of metals which has
been made by Mr. Royce and his assistants in the
Physical Laboratory of this Company. We regard
this as perhaps the most important department in
the works.*

Despite a production run of fewer than these eight
thousand Ghosts, Rolls-Royce had arrived, and to stay.
It had become so famous, so quickly. It was now part
of the canon, the lexicon. It represented motoring's
acme, its exemplar, the sine qua non, the ne plus ultra,
the industry's apotheosis. The *OED* records the car's

eponymous progress through the vocabulary. In 1916, an airplane was described as the "Rolls-Royce of the air." In 1923, there was a reference in the press to a baby carriage being the "Rolls-Royce of the pram-world." Rugs from Isfahan were so described in 1974, pianos by Steinway in 1977, and in 2006, a De'Longhi four-slice, cool-walled, crumb-drawered kitchen device was said to be the "Rolls-Royce of electric toasters." For more than a century now, the agreed-upon name that Sir Charles and Sir Henry chose has become a universal denominator of excellence, its dominance unchallenged, its reputation sealed—and all based on a renown for accuracy, exactitude, and mechanical perfection machined down to the finest and most unforgiving of tolerances.

At about the same time as the Ghost's birth, but four thousand miles away, at a factory in Detroit, Michigan, quite another kind of car was just establishing itself, though it was as different from the handmade paragons of Cooke Street and Derby as it was possible to be. It was the Ford Model T, and it appeared on the roadways of America in October 1908, shortly after the first Silver Ghosts began their wanderings through England and Scotland.

Henry Royce had offered up precision for the few. Henry Ford wanted precision to be available to the many. "I will build a motor car for the great multitude," he declared in 1907. "It will be large enough for the family, but small enough for the individual to run and care for. It will be constructed of the best materials, by the best men to be hired, after the simplest designs that modern engineering can devise. But it will be so low in price that no man making a good salary will be unable to own one—and enjoy with his family the blessing of hours of pleasure in God's great open spaces."

It would be idle to suggest that Henry Ford's early motives were entirely altruistic. He was a Michigan farmer's son who developed an early interest in engineering—and to that extent, the arc of his youth was remarkably similar to that of Henry Royce. Machinery of all kinds quite besotted him. As a teenager, he had become unusually adept, for instance, at repairing his neighbors' pocket watches. His appetite whetted, he then sought and won an apprenticeship—not at a great railway workshop, where Henry Royce was interning at almost the same moment, but at a nearby firm that made very much more mundane objects, such as water valves, steam whistles, fire hydrants, and gongs, and that used an abundance of lathes and drill presses to do so.

He was enraptured by the vision of the majestic steam-powered Westinghouse threshing engines that were occasionally brought in to help his father and other

More than sixteen million Model T Fords—"Tin Lizzies" in the vernacular—were sold between 1908 and 1927, with the price going down from $850 to $260 thanks to the evolution of ever-more-efficient manufacturing techniques.

Henry Ford—like Henry Royce, born into modest circumstances in 1863—went on to popularize motoring and to build the first automotive assembly line in Detroit.

nearby farmers with their harvest, most particularly those that had been designed to propel themselves—the thresher drive belt was removed and looped to power the travel wheels instead. It is a central part of the Ford origin story that young Henry became especially adept at running and repairing a neighbor's portable Westinghouse steam engine, and that, in the summer of 1882, he took a three-dollar-a-day wage to drive this doughty little engine from farm to farm, threshing corn and clover, sawing wood, grinding feed. He fed the engine's fire with old fence posts and corn husks and occasional chunks of coal. Though he found the work backbreakingly hard, Ford claimed never to have been as happy as when he was wandering the dusty Michigan back roads with the Westinghouse, employing simple motive power to help bring farmers some brief contentment, and his teenage self an accumulating wad of paper money.*

---

* Many years later, Ford had his staff searching for a slide valve he had once fashioned for this engine, and which he remembered as being numbered 345. They eventually discovered it broken and abandoned in a Pennsylvania field. To help celebrate his sixtieth birthday, Ford had it repaired and refurbished, fired it up, and used it to thresh corn once again. As to whether "345" was his Rosebud, or whether he simply wanted to remind himself of the design of the slide valve he had made for it, Ford corporate history is unclear.

Before long, he became the demonstrator and repair-man for the local Westinghouse steam engine distribu-tor. Yet, soon thereafter, realizing the one limitation of his beloved threshing engines—no electricity!—he left the world of steam behind to become a mechanical engi-neer at the Edison Illuminating Company, where there was electricity aplenty. It was a precipitate move, but one that, unknown to both, led his life to mimic that of Henry Royce, who was simultaneously learning across the ocean in Manchester what Ford was now learning here in Detroit—the joint and associated worlds of me-chanical and electrical engineering having been brought together since the 1870s in what had been called the internal combustion engine, to produce sustained and efficient motive power.

The parallels continued to be uncanny: for while Royce had bought and tinkered with a De Dion–powered quadricycle as his first-ever vehicle, Henry Ford, who had learned well at his posts with Westinghouse and Edison, made a quadricycle in his spare time, and a two-cylinder petrol engine to power it. It had its first run on June 4, 1896—they had to ax down the workshop door to allow the vehicle out onto the street, as Ford had forgotten how wide he had made its frame—and it soon broke down. Its quickly solved mechanical problem attracted a merry crowd of gawkers, even though the test run was staged after midnight.

And just as with Henry Royce and his progress from De Dion to Decauville and then to his own creations, so Henry Ford progressed swiftly to making his own machines also. There were experiments of one kind and another; there were crude racing cars assembled, engines of two and three and four cylinders; there were successes and failures, setbacks and small victories, quarrels and stutterings, and commercial misery—two early Ford firms failed in less than two years, one of them after making just twenty cars. But by 1903, some sense of stability had settled on the young farmer's son. He had weathered the various crises and was still standing; he knew a great deal about how not to make cars; he now had sufficient confidence in his abilities, sufficient money, sufficient talent, and a sufficient number of friends and admirers to form (because he was able to wrest the name "Ford" from an earlier debacle) the Ford Motor Company, there to begin work on bringing precision engineering to the general public on a prodigious scale.

Yet, while Henry Royce over in Manchester had been captivated by perfection, Henry Ford in Dearborn was consumed by production. Their two fledgling companies, so similar in so many ways, each wedded to the idea of making the best and most suitable machine it could, began to diverge in both purpose and practice from the moment of their respective foundings.

Where Henry Royce began with the Royce Ten, Henry Ford began with the Model A. Like all the early cars Ford would make, the Model A (which was available only in red) was advertised as being "made of few parts, and every part does something." There was no padding, no luxury, no fuss. For additional money, a customer could add features (a rear door, a rubber roof, lamps, a horn, brass trim), but for $750 plus tax, he would get a tiny—the wheelbase was a mere six feet—and inelegantly simple two-seater runabout with an eight-horsepower two-cylinder engine with a semi-automatic transmission that had two forward gears and one reverse, and brakes on the rear wheels only. The little red machine could chug along, reliably unreliable, at a little under thirty miles per hour. Buyers were solemnly warned that a patent infringement case might interrupt their liberating enjoyment of the machine—it never did, and the case, involving a man named Selden, was settled out of court. A Chicago dentist bought the first Model A, and about 1,700 customers followed: Ford was down to its last $223 in working capital by the time the first car sold, so its relative commercial success over the twelve months of its production helped keep the firm afloat, and served as a placeholder for the making of all the subsequent cars that would culminate in Ford's first true success, the phenomenal, society-altering production of the Model T.

Given that the letter *T* is the twentieth letter, one might assume there were eighteen models made after the A; in fact, there were just five: the B (powerful, upscale, costly, with the engine at the front); the C (a fancier A, and like the A, with the engine beneath the seat); the F (a luxury A, sold only in green); the K (a luxury B, but with a six-cylinder engine, also under the hood); and finally, the N (cheap and light, using, for the first time, steel with added vanadium, an alloy that Henry Ford discovered in the wreckage of a crashed French racing car, and which he ordered used as extensively as possible in his future machines, as it gave the chassis added tensile strength and at a markedly lower weight). The Ford Model N cost $500 and had a four-cylinder engine; seven thousand were sold. It was available only in maroon. It was almost the perfect car, thought Henry Ford—except, not quite.

Improvements could still be made, and they were. The Model T was the result. With what came to be known affectionately as the Tin Lizzie, Ford hit the jackpot. The car was officially born on October 1, 1908, and was eventually produced in vast numbers—16,500,000 were sold during the nearly nineteen years of this most stunningly successful model, before the very last Model T rolled off the production line in May 1927.

"Production line" is the key phrase here. All Henry Ford's earlier model cars, just like the Royce Tens and the Rolls-Royce Silver Ghosts across the Atlantic (and later, briefly, in Massachusetts), had been made in the same essential way: pieces, components, parts of the car were all brought to one location on a factory floor, and a jostling gang of men welded and hammered and soldered and bolted and snapped and levered and turned screws and *filed*, always they would *file* to achieve the proper fit, until all the parts came together and—presto!—a new car was shakily and haphazardly born and driven snortingly out into the world.

Then, with the Model T, Henry Ford changed everything. From the start, he was insistent that no filing ever be done in his motor-making factories, because all the parts, components, and pieces he used for the machine would come to him already precisely finished, and to tolerances of cruelly exacting standards such that each would fit exactly without the need for even the most delicate of further adjustment. Once that aspect of his manufacturing system was firmly established, he created a whole new means of assembling the bits and pieces into cars. He demanded a standard of precision for his components that had seldom been either known or achieved before, and he now married this standard to a new system of manufacture seldom tried before. In doing so, he quite transformed industries—his own, but also industries well beyond the making of motorcars—and then, in time, he transformed the world to which industry belonged, for just about everywhere and everybody, and for just about evermore. Though there are other, smaller-scale contenders for the title,* it can fairly be said that in the making of the Model T, Henry Ford cre-

---

* The rudiments of mass production assembly lines, already established at the armories in Springfield and Harpers Ferry (a very American phenomenon, still resisted in Europe and elsewhere), had by this time also been embraced by the New England clock industry and were also revolutionizing the making in particular of three metal consumer products of the times: sewing machines, bicycles, and typewriters. Crucial to all these industries, and absolutely crucial in Henry Ford's new automobile-manufacturing industry, was the use of interchangeable parts. It is worth noting that none of Ford's early-model cars (the A, B, C, F, K, or N) relied entirely on its components' interchangeability. But the Model T did, and did so in spades. Some claim that Ransom Olds was the industrialist who pioneered the use of assembly lines in the making of cars, but he managed to make industrial history confusing by not using interchangeable parts—the workers on his Oldsmobile assembly lines still filed metal pieces to make them fit.

An assembly line, like this at Ford's main plant in Dearborn, Michigan, demands absolute precision in all its components—of which there were fewer than one hundred, compared to some thirty thousand in a modern car. If a single part did not fit, the line would risk being halted—while among the hand-makers at Rolls-Royce, a fit could be achieved with a file.

ated the full-scale and presently recognizable industrial production line.

The Model T had fewer than one hundred different parts (a modern car has more than thirty thousand). How they, no more complex than a modern washing machine, would be assembled into a working automobile was to be Henry Ford's abiding challenge during the first two decades of the twentieth century. For his earlier models, he had experimented with a variety of manufacturing techniques. He had workers in teams of fifteen or so, for example, all bent on building a single car. He then ordered one man to build a car entirely on his own, as from a kit, with other workers bringing all parts and tools to him right as he needed them, like nurses to a surgeon, so that he need never move from his assigned workstation. If there were fifteen such one-man assembly stations on the factory floor, and if the right parts, all precisely made, got to each on time, together with the tools necessary to fit them together, fifteen cars could be made simultaneously in a day.

Then again, in a further experiment, men were assigned a single task for each car-in-the-making, and once having performed that task (bolting on the hood, for example, or fitting the rear bumper), each man would walk to the next car in line and do precisely the same thing again. Parts (hoods, bumpers, cylinder blocks, lights) were made by much the same means, on an upper floor of the three-story plant, and were stored upstairs and sent down chutes to the assembly floor, meaning that there were never any Himalayas of parts to impede the workers' progress on the floor, and yet there were always freshly made components available on tap, as it were.

Each of these systems had its advantages; each represented another accretion of manufacturing knowledge and wisdom—until 1913, when there came at last the *Caramba!* moment—the discovery that *the workpiece could be moved along* in front of the workers, who would each perform a single very ordinary and undemanding task on it as it presented itself, and would then do the same for the next, and the next, and the next as each passed briefly before them, while other workers performed very different other tasks again and again and again as the workpiece presented itself before each of *them*, until a whole new piece was made, a whole new assembly was fashioned, and out of these various assemblies and assemblages, a whole new car was manufactured, by the accumulation of hundreds or thousands of one-task-at-a-time performances as the car-to-be moved along what would be, in effect, the entire length of its automotive birth canal.

Henry Ford said he first got the idea for the assembly line from watching pig carcasses being solemnly and meticulously disassembled in a local pork butchery: one had merely to reverse the process of slicing and boning and draining and rendering and deconstructing, and instead, weld and bolt and bronze and construct and then spray with paint (quick-drying black, the only choice). And where, on the one hand, you had chops and hams and chitlins and grease, here at the new Ford plant, you fashioned out of metal and glass and rubber parts a brand-new car, to be sold for eight hundred dollars and change.

And the speed of the new way! The revolutionary productivity! The first device to be made in this assembly-line manner was the Model T's magneto, the simple magnet-and-two-coils device that would produce the spark for the ignition of the fuel in the engine. At the factory, Ford made a long, straight line at waist level with a conveyor belt, and on it, at first, just the already-made simple steel wheel that would be turned by the car's crank handle. The first person in the line, sitting before the conveyor belt and so being confronted with the wheel moving steadily across his or her field of vision, would bolt to it a small electric coil that had been prewound with maybe two hundred turns of copper wire. The next person sitting in line would bolt on a smaller wheel, but this time a wheel wound with perhaps two thousand turns of a much finer copper wire. The third person would attach to the wheel a case with a U-shaped magnet attached to its inside, and then a fourth person would bolt the cases together and send the finished magneto for inspection.

Someone, a tester, would spin the coils through the magnetic field: a weak electric current would be induced in the two-hundred-turn coil, then a very much larger voltage would be created in the two-thousand-turn coil, and if all was working and the parts were made as precisely as specified and fitted together as they should, a mighty spark would flash between the terminals at the magneto's business end—which, had this been mounted on an engine and not on the assembly line for a test, would have flashed at the top of the cylinder just at the moment when it was filling with a highly flammable mixture of vaporized gasoline and air, and there would be an explosion that would thrust the piston downward and set the Ford's powerful little motor off and running.

Before the advent of the assembly line, it would take a worker twenty minutes to assemble a magneto from scratch. With the assembly line running and its gang of workers each performing a single mind-numbing task, it would take just five minutes to make a complete magneto—and each magneto would be identical, none subject to a worker's whim or a bout of Friday laziness, and all would fit in their assigned location inside the Ford's engine, and would sit comfortably on location without a scintilla of doubt.

Axles were the next car parts to be assembled on a line, an axle line being put into operation sometime in 1915. It used to take two and a half hours to put one axle together; on the new line, twenty-six minutes. Another production line then cut by exactly half the time needed to put together a transmission, with its three forward and one reverse gears arranged in Ford's curious plan-

THE IRRESISTIBLE LURE OF THE HIGHWAY · 255

etary system of belts and slipping wheels. Where once it had taken workers ten hours to construct an entire engine, now it took just four—helped by a brand-new Ford design for a cylinder block, with its top and bottom shaved off to accommodate the valves and plugs above and the crankshaft and lubricant sumps below, and with the cylinders themselves now easily accessible to the machine tools that would bore the holes of such infinitely exact depth and diameter. In time, a new Model T would appear at the factory doors in Dearborn every forty seconds.

Almost no skill was needed to work on an assembly line, whereas, for an engineer to measure tolerances and file for fit and test and retest and employ go and no-go devices—all this did take craftsmanship, training, extra pay. Henry Ford found he had managed to solve a host of problems like this in one fell swoop. By creating assembly lines, he could produce countless more cars; he could do so inexpensively; he could charge lower and lower prices and make his vehicles more and more affordable, popular, and ubiquitous; he could employ people of less and less skill to make them; and he could do away with the need for craftsmanship, leaving that to people who made Rolls-Royces.

The personal result was that while Henry Royce

became respectably well off as a consequence of his endeavors, Henry Ford, by contrast, became one of the wealthiest men on the planet and in all that planet's history—and he left as legacy not just a car company that to this day remains one of the world's largest, but a foundation that spreads his legacy of wealth to the deserving many around the world.

And precision's differing roles in the two companies? Within Rolls-Royce, it may seem as though the worship of the precise was entirely central to the making of these enormously comfortable, stylish, swift, and comprehensively memorable cars. In fact, it was far more crucial to the making of the less costly, less complex, less remembered machines that poured from the Ford plants around the world. And for a simple reason: the production lines required a limitless supply of parts that were exactly interchangeable. If one happened not to be so exact, and if an assembly-line worker tried to fit this inexact and imprecise component into a passing workpiece and it refused to fit and the worker tried to make it fit, and wrestled with it—then, just like Charlie Chaplin's assembly-line worker in *Modern Times* or, less amusingly, one in Fritz Lang's *Metropolis*, the line would slow and falter and eventually stop, and workers for yards around would find their work disrupted, and parts being fed into the system would create unwieldy piles, and the supply chain would clog, and the entire production would slow and falter and maybe even grind, quite literally, to a painful halt.

Precision, in other words, is an absolute essential for keeping the unforgiving tyranny of a production line going. As far as a handmade car is concerned, though, upfront precision is quite optional. It is a need that could be attended to during the hand-making process, as the process itself never depends (at least, not in the Silver Ghost days) upon every component's being precise from the commencement of manufacturing. The irony remains: a Rolls-Royce is so costly and exclusive and has enjoyed for so long a reputation of peerless creation and impeccable performance, but it does not require absolute precision at all stages of its making. A Model T Ford, however (or, indeed, any modern car, now made by robots rather than human beings, by Chaplinesque figures staring glassy-eyed at the endlessly flowing river of parts), requires precision as an absolute essential. Without it, the car doesn't get made.

There is one further component to this story: the use by Henry Ford of an invention that helped make it possible for the cost of his Model T to decrease almost every year during its eighteen years of production, to go down in price from $850 in 1908 to $345 in 1916, to a stunningly affordable $260 in 1925.

The car was the same, the materials the same, but the means of production had become vastly more efficient. Henry Ford had been helped in his aim of making it so by using one component (and then buying the firm that made it), a component whose creation, by a Swedish man of great modesty, turned out to be of profoundly lasting importance to the world of precision.

The Swede was Carl Edvard Johansson, popularly and proudly known by every knowledgeable Swede today as the world's Master of Measurement. He was the inventor of the set of precise pieces of perfectly flat, hardened steel known to this day as gauge blocks, slip gauges, or, to his honor and in his memory, as Johansson gauges, or quite simply, Jo blocks—the same polished steel blocks and tiny billets my father brought home to show me back in the mid-1950s as an example of what precision was truly all about.

Carl Edvard Johansson got the idea while on a train. He was at the time, in 1896, working as an armorer-inspector at a government-run firearms factory in the city of Eskilstuna, Sweden's steel-making equivalent of Pittsburgh or Sheffield, and which still has a steelworker on its coat of arms. His plant had been making Remington rifles under license but was just then switching to a variant of the German Mauser carbine and, in the process, was changing to an entirely new system of measuring. Johansson, who had an abiding respect for ultra-precise measurement, had gone to the Mauser factory in the German Black Forest to investigate the company's ways of measuring, and for some reason, he found its scheme wanting. According to legend, he was pondering the idea of making improvements to the forthcoming Swedish operation while on the long and otherwise tedious rail journey home.

His idea was to create a set of gauge blocks that, if held together in combination, could in theory measure any needed dimension. What, he wondered, was the minimum number of blocks that would be needed, and what should the sizes of the various blocks be? By the time he stepped off the clanking steam train at Eskilstuna station, he had solved the problem: with just 103 blocks made of certain carefully specified sizes, arranged in three series, it should be possible, he said, to take some twenty thousand measurements in increments of one one-thousandth of a millimeter, just by laying two or more blocks together.

It took Johansson some long while to make the first prototype set—he used his wife's sewing machine, converting it by adding a grinding wheel, to smooth the blocks to their correct dimensions. It was a task well suited to his personality, a biographer later recalled. For Johansson was, by all accounts, a modest, retiring, unassuming, private, pipe-smoking, mustachioed, patient, formal, stooped, eternally avuncular son of the croft, a man who grew up on a rye farm in central Sweden and, yet, went on to change the world. The 103-piece combination gauge block set he eventually developed, according to his biographer, has since "directly and indirectly taught engineers, foremen and mechanics to treat tools with care, and at the same time giving them familiarity with [dimensions of] thousandths and ten thousandths of a millimeter."

Gauge blocks first came to the United States in 1908, the initial set of them brought through customs by Henry Leland, the machinist and precision fanatic best known as the Man Who Invented the Cadillac.* Just as with the nineteenth-century demand for wooden pulley blocks for the Royal Navy—no connection at all, other than ironically—sales of the new Jo blocks rocketed, as more and more industries were established, all of them demanding this simple and elegant means of measuring

---

* As well as the Lincoln—and the electric starter motor, which he built after his best friend was knocked out and killed by the unexpected kickback of a large car's starting crank.

Henry Ford bought the American gauge block business of its inventor, Carl Edvard Johansson, the Swede still known today as the world's "Master of Measurement." With the use of so-called Jo blocks, extreme tolerances could be realized swiftly, further increasing the efficiency and reliability of engineered products.

their various products. Eventually, Johansson himself was persuaded to set up shop in America, first in New York and then to make block sets in an old three-story piano factory in Poughkeepsie, a hundred miles to the north, on the Hudson River. His arrival was greeted by the press: "The Most Accurate Man in the World," said one. "The Edison of Sweden."

At the time, Henry Ford did not make use of Jo blocks in his factories, even though his entire system of mass production depended wholly on the most extreme accuracy. Whether he was implacably opposed, or whether there was some other reason, remains unclear: his opposition or insouciance ended swiftly, however, once he became aware of a sharp exchange between his factory managers and the Swedish ball-bearing maker SKF.

This firm, founded in 1907 and still in existence—its initials stand for Svenska Kullagerfabriken AB—was receiving from Ford in the 1920s what it claimed were numerous "unjustified complaints" regarding the dimensions of its bearings. Ford workers on the Detroit production lines claimed that the SKF bearings were often significantly out of true, and were causing delays and stoppages on the factory floor. SKF managers protested robustly, insisting that their bearings were perfectly spherical, and that measuring them using Jo blocks would demonstrate that this was so.

As indeed the Jo blocks duly demonstrated. If any complaints were to be leveled, said officials from SKF, they should by rights be leveled at the machines and assembly lines on which the bearings were being used—and Henry Ford, to his horror, realized that they were right. Maybe, he said to his colleagues assembled for an emergency meeting, his cars were precise only to *themselves*; maybe every manufactured piece fit impeccably because it was interchangeable to *itself*, but once another absolutely impeccably manufactured, gauge-block-confirmed piece from another company (a ball bearing from SKF, say) was introduced into the Ford system, then maybe its absolute perfection trumped that of Ford's, and Ford was wrong—ever so slightly maybe, but wrong nonetheless.

So Ford, being powerful and rich and unstoppably ambitious, did what others might not have had the moxie to do. He made contact with Johansson and persuaded him to move his entire gauge block production process seven hundred miles, from Poughkeepsie to Detroit, and set up shop within the vast new Ford factory there. Johansson did as he was bidden, and in due course, and in line with Ford's relentlessly persuasive manner, he then sold up and allowed his small, elderly, old-fashioned, yet vitally important business to become a division of the Ford Motor Company—to be swallowed up, in other words—and then, in 1936, left Henry Ford to his own devices and went quietly back to his native Sweden, there to collect gold medals and honorary degrees and visiting fellowships and royally bestowed distinctions in impressive numbers.

Johansson grew deaf in his later years, and used an ear trumpet, which he called his pipe of peace. He once met Edison, who was deaf as well, and he liked to recall how the two great inventors put their heads together, quite literally, and discussed gauge blocks, which by this time, after the Great War, were achieving accuracies of up to one-millionth of an inch. But can you maybe do better than even that? Edison inquired. Yes, replied Johansson, it was now possible to achieve precision tolerances down to one ten-millionth of an inch, but he would not reveal exactly how. Quite right, the notoriously cantankerous and ungenerous Edison harrumphed. Far better to keep quiet where matters of invention are concerned.

Carl Edvard Johansson died in 1943, respected and beloved in Sweden, and forgotten elsewhere. The industrial system of mass production that his invention unwittingly helped refine and expand, and which relies on as absolute a degree of precision as is attainable, continues to this day—on the ground and, more perilously, high up in the air as well, where any failure of precision can be unimaginably dangerous.

Chapter 6

(Tolerance: 0.000 000 000 001)

Precision and Peril, Six Miles High

It was like love at first sight: [Frank] Whittle held all the winning cards: imagination, ability, enthusiasm, determination, respect for science, and practical experience—all at the service of a stunningly simple idea: 2000 hp with one moving part.

—LANCELOT LAW WHYTE, "Whittle and the Jet Adventure," *Harper's Magazine* (January 1954)

When it comes to the steady and uncomplaining workings of a device such as a tricycle, a sewing machine, a wristwatch, or a water pump, mechanical perfection is naturally a good thing—but perhaps it is also a thing that is seldom essential to the preservation of life and limb. In the matter of a high-powered sports car or an elevator or a robotic operating theater, however, precision comes to be a vital necessity: mechanical failure occasioned by imprecision at a hundred miles an hour or on the sixtieth floor of a skyscraper or in the middle of heart surgery could have terrible, maybe lethal results.

Furthermore, in those situations where high speed is combined with high altitude, when paying customers are suspended unnaturally many miles above the planet's all-too-hard surfaces, and, moreover, in a place where the human presence is inherently unwelcome and life unsustainable, the precision of the aircraft machinery that took them up there has to be utterly impeccable. Any departure from absolute perfection could have the potential for the gravest and most disastrous of consequences—as the world came to know just a few minutes after 10:00 a.m. on the sunny Singapore morning of Thursday, November 4, 2010.

Qantas Flight 32, a two-year-old Airbus A380 double-decker "superjumbo" jet aircraft, at the time the largest commercial airliner in the world, was beginning a routine seven-hour flight down to Sydney. There were four hundred forty passengers aboard, two dozen cabin crew, and a slightly unusually large number of five men in the cockpit: a captain, a first officer, a second officer, a check captain, and a supervising check captain, this last on board to check the check captain, who in turn was there to check the performance of the rest of the crew. Among them, the five had an accumulated total of seventy-two thousand hours of flight time, an amassment of experience that would be sorely needed that morning.

The aircraft had taken off at two minutes before ten from one of Changi Airport's two southwest-heading runways, 20C. The plane's landing gear was promptly retracted, the thrust settings on the four Rolls-Royce Trent 900-series engines were set to Climb mode, and the 511 tons of aircraft, cargo, and human passengers began to claw their relentless way upward. Within moments, the aircraft had left Singaporean airspace and entered that of the Republic of Indonesia. It was powering into the cloudless sky at a mile and a half above the mangrove swamps and small fishing villages of Batam Island when, suddenly, to the surprise, dismay, and consternation of almost everyone aboard, there were two very loud bangs, one quickly after the other.

The captain immediately overrode the automatic pilot and ordered his aircraft to cease climbing, to keep itself level at seven thousand feet, and to maintain its southerly heading. The cockpit monitors at first indicated only one event: the overheating of a turbine in the number two engine, which was on the left wing, the inboard engine, the one closer to the fuselage. Within seconds, though, this single announcement became a drizzle, then a flurry, and finally a violent storm of flashing lights and sirens and alarm bells as, one after another, systems all around the aircraft were shown to be failing. And within the number two engine, the overheating had now transformed itself into a raging fire.

The captain radioed a "pan-pan" message back to Singapore air traffic control, a message indicating a serious problem, though less than an all-out emergency. He then decided to turn back toward Singapore, to ease himself into a racetrack holding pattern, to use half an hour of stable flying to work out what exactly had happened to the engine, and to decide how best to deal with the cascade of problems its breakdown had occasioned. Meanwhile, fuel could be seen gushing from the rear of the engine, and a peppering of holes could be seen in the wing, where debris from an explosion of some kind had clearly smashed into it. Reports were also coming in from down below that pieces of aircraft engine had been found in villages on Batam Island, all of them clearly spewed from the damaged plane.

Takeoff is optional, they say; landings are compulsory. It took the better part of an hour for the crew to deal with the various problems afflicting their stricken aircraft, and to work out just how to land when all manner of critical parts of the aircraft were no longer working. The brakes, for example, seemed to be only partially functioning, the spoilers on the left wing could not be deployed, there was no working thrust reverser on the failed engine, and the landing gear could not be properly cranked down for touchdown. The plane would thus come in for a very fast landing, and with ninety-five extra tons of fuel aboard and badly compromised brakes, it might not be able to stop itself before running out of the almost three miles of runway. The airport was asked to scramble its fleet of emergency vehicles and wait for the approach of the giant jet.

In the event, the massive plane did manage to bring itself to a stop—the captain near-frantically pressing the pedals hard to the metal—with just over four hundred feet of runway left. What didn't stop was the number one engine, on the left wing's outboard. The number two had been fatally damaged and was not running, but for some reason—because the control cables and electrical connections had been severed, it later transpired, by whatever had crashed through the wings—its near neighbor still was.

Moreover, torrents of fuel were still gushing from

ruptured tanks near the number two engine, and most worrisome of all, such brakes as remained on the left-hand side of the aircraft body had overheated during the high-speed, heavy landing and were now red-hot, registering almost a thousand degrees Celsius on the cockpit display.

To add to the grisly picture, the tires had ruptured and were ripped and flat, allowing the bare metal of the wheel rims to scrape along the runway for hundreds of feet. Were any wafts of the gushing fog of fuel, perhaps blown by the jet thrust from the unshutdownable number one engine, to have washed over the near-incandescent brakes or the superhot wheel rims, there would likely have been a spark, a sudden flash of fire, and then, when the wing fuel tanks were properly heated, an almighty explosion. The brief relief of the safe landing would have been replaced by the utter horror of an immobilized plane fully consumed by flames. It was a chaotic and terrifying situation—now much worse on the ground, it seemed, than ever it had been up in the air.

It took the Singapore firefighters three hours to stop the running engine, in effect by drowning it with high-powered jets of thousands of gallons of water. Engines are built to withstand rainstorms, and it is a testimony to the robust design and construction of these

Rolls-Royce Trents that it took so immense a simulated rainstorm to bring this fast-spinning machine to a stop. But just as soon as it became evident that the engine would be brought under control, and once the thousands of pounds of fire-retardant foam and dry-powder fire suppressor had brought the red-hot brakes back to black and reasonable cool, the passengers, who had been cooped up for two hours in what seemed like a firetrap, were let out, clambering down a set of stairs brought to the seldom-used right-hand doors. Many of the four hundred forty were terrified, but no one was hurt.

And then the crew was finally able to see what had happened. It was an ugly sight, seldom seen or experienced by even the most senior of the flight crew. They could now see that the aft third of the cowling of the number two engine had been torn away, the turbine section of the engine stripped naked, and two gaping holes were visible where part of the engine structure had been blown apart. There was soot, oil, burned wiring, smashed pipes, and parts of damaged rotor blades everywhere.

A heavy metal rotor disc, it turned out, had burst out of the engine; about half of it, torn into hot fragments, was to be found down in the villages of Batam Island. The fragments had rained down from the plane, smashing onto buildings but hitting no one.

What had happened was the nightmare of every jet engine manufacturer in the world. The Rolls-Royce Trent 900—specifically a 972–84 variant, which developed almost seventy thousand pounds of thrust and had cost Qantas Airways $13 million—had suffered what is known as an in-flight uncontained engine rotor failure. This is an exceptionally rare occurrence, but when it happens, it is invariably an exceptionally violent one, with hot metallic components from the engine fracturing and, rather than being enclosed by the metal casing, tearing through it and then being thrown out as shrapnel to tear through the wings and the fuselage of the aircraft.

Bundles of electrical cables, fuel tanks, fuel and oil pipes, hydraulic systems, mechanical systems, and a pressurized passenger compartment with highly vulnerable human bodies within—all these can be hit and damaged by ragged chunks of fast-flying metal. In the case of Qantas Flight 32, many were, and a tidal wave of destruction ricocheted through the aircraft. To the relief of all concerned, the damage and loss of control were successfully managed by a highly competent (and unusually numerous) crew on the flight deck.

But exactly what had taken place inside the engine to bring about this near catastrophe? To appreciate that, and to enter the ultraprecise but still Hadean nightmare that is the interior of a modern jet engine, requires some history—and a return to the time, not so long past, when aviation was a propeller-driven pursuit still available to the enthusiastic amateur rather than the digitized zealotry found in today's commercial airline cockpits.

It was Frank Whittle, the first son of a Lancashire cotton factory worker turned tinkerer, who invented the jet engine. There were other contenders, although for the kind of engine most widely recognized nowadays—

The wholly destroyed Rolls-Royce Trent number two engine after the safe landing in Singapore of Qantas Flight 32 following an "uncontained rotor failure" inside the engine a mile high over Indonesia.

Frank Whittle conceived the basic idea of a jet engine while still a young flying student, though for want of the fee was unable to renew the patent. His first propeller-less jet plane flew in May 1941.

the air-breathing internal combustion engine that powers most jet aircraft today, and manifestly not the non-air-breathing rocket, which is technically a jet engine also—there are really only two. One was the Frenchman Maxime Guillaume, who secured the French government's *brevet d'invention* Number 534,801 for a turbojet aero engine in April 1922; the other was Hans von Ohain, from Dessau in Saxony, who, in 1933, came up with what he felt certain was a workable design for "an engine which did not require a propeller," and actually saw it made.

Yet neither the French idea nor the German prototype flourished. The technical requirements for an engine that was destined to function in environments of extreme physical hostility, particularly with such a fierce predicted heat that would envelop all its working parts, were just too daunting for both the materials and the engineering skills available in Europe at the time. Also, it is worth noting that American laboratories were curiously blind and deaf to the idea of a turbine-powered engine as having any utility for the aircraft industry, and the United States pursued almost no research until the 1940s.

It was left to the diminutive Frank Whittle, therefore, to pursue the dream, fired by his famous criticism of the outmoded nature of propeller-driving piston engines, a condemnation that resonates today. "Reciprocating engines are exhausted," he declared. "They have hundreds of parts jerking to and fro and they cannot be made more powerful without becoming too complicated.* The engine of the future must produce two

---

* Whittle, only half joking, blamed his prejudice against piston engines on his catalog of motorcycle accidents, which culminated in his failure to stop at a T junction outside London and being catapulted into the woods, and then having his insurance canceled and his ruined bike repossessed by the finance company. Whittle was not a man to blame himself for mishaps, and instead

thousand horsepower with one moving part: a spinning turbine and compressor."

Modern jet engines can produce more than a hundred thousand horsepower—still, essentially, they have only a single moving part: a spindle, a rotor, which is induced to spin and, in doing so, causes many pieces of high-precision metal to spin with it. Jet engines are beasts of extreme complexity bound up within a design of extraordinary simplicity. All that ensures they work as well as they do are the rare and costly materials from which they are made, the protection of the integrity of the pieces machined from these materials, and the superfine tolerances of the manufacture of every part of which they are composed. Frank Whittle had to deal with these harsh realities for ten testing years, from the moment he had his grand idea in the summer of 1928. Every imaginable obstacle was put in his way during that decade. Nevertheless, he persisted.

Frank Whittle, five feet tall, slightly Chaplinesque in appearance, neat, punctilious, and seemingly made of compressed steel springs—as a youngster, he was a daredevil stunt flier and demon motorcyclist, an irritant to his instructors, and a mathematician of rare

---

blamed the motorcycle engine for taking him so uncontrollably fast.

ability—first planted the seed at the end of his stint as a flight cadet at Cranwell, the Royal Air Force academy in the English Midlands. Cadets at the time were each obliged to write a short scientific thesis on a topic that interested them, and Whittle's paper has since become a part of aeronautical legend: with all the hubris of a young man on the make, he titled it "Future Developments in Aircraft Design."

At the time of his graduation from Cranwell, powered flight was only a quarter of a century old. The aircraft in which cadets such as Whittle trained were mostly biplanes—they had wooden frames, were in no sense streamlined, had no enhancements such as retractable undercarriages or pressurized cabins, flew at low altitudes, and trundled through the skies at speeds seldom exceeding 200 miles an hour. RAF fighters, in many ways more advanced than most, averaged a puny 150 miles per hour, and operated at only a few thousand feet above sea level.

Science fiction was the reading rage of the day, and to a reader such as Whittle, who consumed all the H. G. Wells and Jules Verne and Hugo Gernsback he could lay his hands on, the fantastical possibilities on offer (of high-speed flight, of mass transportation, of journeying in the stratosphere, to the moon, to outer space!) presented not just a contrast, but also, in his considered view, an *achievable* contrast. Whittle believed that all that the fantasists were offering could actually be achieved, and yet not, he insisted, with the reciprocating engines of the moment. A new and better kind of engine was needed. He later described the ideas he advanced in his memorable Cranwell thesis.

> *I came to the general conclusion that if very high speeds were to be combined with long range, it would be necessary to fly at great height, where low air density would greatly reduce resistance in proportion to speed. I was thinking in terms of a speed of 500mph in the stratosphere where the air density was less than one quarter of its sea-level value.*
>
> *It seemed to me unlikely that the conventional piston engine and propeller combination would meet the power plant need of the kind of high speed/high altitude aircraft I had in mind, and so in my discussion of power plant I cast my net very wide and discussed the possibilities of rocket propulsion and of gas turbines driving propellers, but it did not then occur to me to use the gas turbine for jet propulsion.*

It was fifteen months later, in October 1929, that the penny finally dropped. Whittle was by now a fully qualified general duties pilot, stationed in Cambridgeshire, and while training and teaching others to fly, he obsessively ruminated and calculated and imagined the kind of engine that could possibly make aircraft go lightning fast. All his designs involved some kind of supercharged piston engine. At the same time, he could see that even a modest increase in engine power, and thus aircraft speed, would require a very much larger and heavier engine—an engine probably much too big and heavy for any aircraft to carry. He was about to abandon the quest when, suddenly, one day that October, he had his brain wave: why not, he thought, employ a gas turbine as an engine, a gas turbine that, instead of driving a propeller at the engine's front, would thrust out a powerful jet of air from the engine's rear? An idea that would change the world in unimaginable ways had come to Frank Whittle when he was just twenty-two years old.

His recent school days, and his later mathematical skills, reminded him that a propelling jet of the kind he was proposing would offer a working demonstration of Isaac Newton's Third Law of Motion, propounded back in 1686. Newton (a Cambridge man, as it happens) had written that "for every force acting on a body, there is an equal and opposite reaction." Under this law, a powerful jet being thrust from the rear of an aircraft engine would drive that aircraft forward with equal power and, in theory, at almost any imaginable speed.

Moreover, a gas turbine could also, in theory, be vastly more powerful than a piston engine, and for a very simple reason. A crucial element in any combustion engine is air—air is drawn into the engine, mixed with fuel, and then burns or explodes. The thermal energy from that event is turned into kinetic energy, and the engine's moving parts are thereby powered. But the amount of air sucked into a piston engine is limited by, among other factors, the size of the cylinders. In a gas turbine, there is almost no limit: a gigantic fan at the opening of such an engine can swallow vastly more air than can be taken into a piston engine—as a rule of thumb, seventy times as much, in the early, Whittle-era jets. Seventy times as much air might not mean seventy times as much power—other factors come into play—but a good twenty times as much power is a reasonable and accepted figure.

Small wonder, then, that this was for the historians of inventions, of breakthroughs, a true eureka moment. It truly did represent, cliché though it may sound, a paradigm shift. And from that autumn day onward, Frank Whittle thought of little else than getting a gas turbine to work sufficiently well to propel an airplane, solving as he did so the endless raft of problems, technical and official, that kept the project from its immediate resolution. It would be ten years before the first working engine was fired up, and as so often happens, it was war that provided the spur.

Not at first, though. Few seemed interested: even though Whittle managed to apply for (and eventually be granted, in 1931) a patent for "Improvements Relating to the Propulsion of Aircraft and Other Vehicles," and even though his officer colleagues at the air base spread the word that he was onto something remarkably innovative and original, he was rebuffed at every turn. The Air Ministry in particular said it had no interest, the three principal British makers of aero engines turned him away, and when, in 1935, it was time to renew his patent, he could not afford the five-pound fee—and the Air Ministry said in no uncertain terms it would not foot the bill out of government funds. Whittle was by then on the verge of giving up, and had developed plans for another kind of device altogether, one that had relevance not to air transport but to journeying by road. He let his cherished patent lapse—not, he thought grimly, that it had any residual value—except that, with the patent's lapse, and with the release of the idea from his exclusive ownership, the world now had access to it, and that had consequences.

For, in 1935, Germany's now-fast militarizing—and the interest in jet engines expressed by Hans von Ohain and the Heinkel Company and coincident fresh enthusiasm for turbine propulsion from the head of airframes at the Junkers factory, Herbert Wagner—placed it firmly in a position to develop a turbojet. Whether either man's interest was spawned by the freeing of the Whittle patent has never been fully established, but the result was self-evident: come the mid-1930s, Germany had indeed become officially interested in producing an aircraft jet engine, while Britain, despite having the idea's patented creator living with his new family and no money and no support for his ideas not fifty miles from the capital, and employed by no less than the Royal Air Force, was not.

This would all change once money was pumped into the project, and Whittle could begin translating his blueprints into test-bed engines and, eventually, see if his ideas would fly, literally. It was a firm of venture capitalists named O. T. Falk and Partners that eventually, in 1935, took the gamble. "Stratosphere plane?" was the note taken on September 11 of that year by the firm's senior partner, Lancelot Law Whyte, who confessed to "falling in love at first sight" with the young officer. Despite the query of his notation, he later told his wife that the experience of first meeting Whittle (who by now was pursuing a doctorate at Cambridge, while on leave from the RAF) was akin to "meeting a saint in an earlier religious epoch." Had one not known the end of the story, it might be easy to suppose that, with a beginning like this, all would inevitably end in tears. Far from it. It ends triumphantly, with the saint indeed performing all the miracles expected of him. And Lancelot Law Whyte emerges from the story as a visionary, a man undeservingly forgotten. He had once been a physicist; he was anything but a coldhearted banker, but was an almost mystical figure, who loved Whittle's idea not because it might make money, but because of its sheer elegance, and because "every great advance replaces traditional complexities by a new simplicity. Here it was in the iron world of engineering."

The firm offered a three-thousand-pound advance and established a company for Whittle, to be named Power Jets Limited. There was little by way of aviation experience among the principals—one of the main shareholders made cigarette vending machines—but Frank Whittle was made chief engineer, and the company's only employee. The Air Ministry (his usual employer, as he was a serving air force officer) agreed for him to be briefly separated from his military duties, though noting that his work for Power Jets was to be spare-time employment only, with his devoting no more than six hours a week to this newfangled idea.

The ministry's support may have been given grudgingly, but it was nonetheless given,* and it was this official "Oh, alright then" backing that convinced Whyte to get going. He placed an immediate contract with the turbine makers British Thomson-Houston[†] to develop an engine to Whittle's specifications. It was to have a turbine spinning at 17,750 rpm that would drive a compressor and produce 500 hp, with the outflowing air creating enough propulsive power to fly a small mail-delivery airplane. It would be called the WU, or "Whittle Unit." Whittle envisaged it as fast enough to

---

* As might be expected, high panjandrums in the prewar scientific civil service took differing views of Whittle's proposals: a man named Harry Wimperis took against the idea—"Many had burned their fingers on gas-turbine projects and I don't suppose you will be the last," he remarked caustically to a Power Jets investor—but Wimperis's senior, the legendary Henry Tizard, was very much a supporter, and ultimately it was Tizard's view that prevailed. Tizard and Wimperis famously later worked together on the invention of radar. If Wimperis was skeptical of jet propulsion, it was but a temporary aberration: he was more generally an open-minded figure, as befits a recipient of a Whitworth Scholarship at Cambridge, named for the great Victorian engineer who was central to the story of precision a hundred years before.

[†] Whittle had approached BTH with his ideas some years before, but it turned him down flat. Now that he had financial backing, BTH was persuaded to take a chance and build a prototype.

carry some few tons of mail across the Atlantic Ocean, nonstop, in about six hours.

At the distance of eighty years, it is scarcely possible to appreciate the revolutionary novelty of this idea. This was no invention that was happened upon by chance. This was a well-planned, carefully thought-out, and diligently evaluated creation of an entirely new means of propulsion, of transportation. This was the moment (or the invention, or the personality) that took the standard model of precision and transported it from the purely mechanical world into the ethereal. What was about to be constructed was a device of transcendental beauty, and though it might be said that mankind has taken the invention of the jet engine and quite spoiled the world with it, the thing itself had then and has still elegance and integrity like few other modern creations.

The basic principle of the turbine engine was already well established, and turbines were already being made (not just by firms such as British Thomson-Houston, but all over the world). Gas turbines were already beginning to power ships, to generate electricity, to run factories. The simplicity of the basic idea was immensely attractive. Air was drawn in through a cavernous doorway at the front of the engine and immediately compressed, and made hot in the process, and was then mixed with fuel, and ignited.

It was the resulting ferociously hot, tightly compressed, and controlled explosion that then drove the turbine, which spun its blades and then performed two functions. It used some of its power to drive the aforementioned compressor, which sucked in and squeezed the air, but it then had a very considerable fraction of its power left, and so was available to do other things, such as turn the propeller of a ship, or turn a generator of electricity, or turn the driving wheels of a railway locomotive, or provide the power for a thousand machines in a factory and keep them running, tirelessly. Chemical energy, produced by mixing air and fuel, was thus transformed into mechanical energy. Mechanical energy was often just what was needed, to drive a ship or a factory, but if it then drove a generator, there was another level of transformation, of mechanical energy being transformed into electrical energy.

Frank Whittle was interested only in the transformation of chemical into mechanical energy. Electricity was to him of only peripheral interest. Yet he wanted the mechanical energy not simply to drive a spinning shaft. He wanted it to create a propulsive jet of gas—and further, he wanted the device that changed chemical energy into this propulsive jet to be light enough to be carried aloft, and efficient enough to make a jet engine good sense economically. This meant that the engine components had to be made with the greatest of care, to very exacting standards, and to be allowed to operate in the harshest of environments. This is what Power Jets and BTH set out to do, starting in 1936. It proved technically difficult in the extreme, and just when Hitler was starting to breathe down everyone's neck.

Heat was probably the trickiest problem. The engine's combustion chamber would create temperatures quite unknown to anyone then involved with burners and boiler-firing equipment. Bearings, too, presented problems—no one had ever invented a bearing that would do its work in the kind of temperatures and pressures likely to be encountered in the beating heart of a jet engine. And the experiments that BTH had to perform—testing fires at all kind of temperatures, testing bearings to their breaking point, producing billowing fumes and dangerous lakes of fuel, and explosions, explosions all the while—no one could explain what was going on, because all was classified top-secret.

It was perhaps just as well, as the first test runs of a completed engine were a near-total disaster. They took place in April 1937, with the plant outside the town of Rugby, in the English Midlands, well prepared for catastrophe—if turbine pieces fracture and get thrown out of an engine, they can be lethal. In an incident some weeks prior, a conventional turbine had exploded and hurled chunks of red-hot metal two miles away, killing several people en route. So the test engine was mounted on a truck (which, because its starter motor weighed a couple of tons, had to have its wheels removed) and was shielded by three pieces of inch-thick steel. Its jet pipe was routed out of a window, and the control for the starter motor was several yards away, with Whittle giving his orders by hand signal to the brave or foolhardy fitter employed to work it.

Whittle's report was not exactly the cool and laconic writing of an experienced test pilot:

I had the fuel pump switched on. One of the test hands then engaged the starter coupling (which was designed to disengage as soon as the main rotor of the engine over-ran the starting motor) and I gave hand signals to the man on the starter control panel.

The starter motor began to turn over. When the speed reached about 1000 rpm I opened the control valve which admitted fuel to a pilot burner in the combustion chamber, and rapidly turned the handle of the hand magneto to ignite the finely atomized spray of fuel which this burner emitted. An observer, peering through a quartz observation window in the combustion chamber, gave me the "thumbs up" sign to show the pilot flame was alight.

I signalled for an increase of speed for the starter motor, and as the tachometer indicated 2000 rpm I opened the main fuel control valve.

For a second or two the speed of the engine increased slowly and then, with a rising shriek like an air-raid siren, the speed began to rise rapidly, and large patches of red heat became visible on the combustion chamber casing. The engine was obviously out of control. All the BTH personnel, realising what this meant, went down to the factory at high speed in varying directions. A few of them took refuge in nearby large steam engine exhaust casings, which made useful shelters.

I screwed down the control valve immediately, but this had no effect and the speed continued to rise, but fortunately the acceleration ceased at about 8000 rpm and slowly the revs dropped again. Needless to say this incident did not do my nervous system any good at all. I have rarely been so frightened.

It happened again the next day, sheets of flame spouting from the jet pipe, vapor from leaking joints being ignited by red-hot metal in the combustion chamber, flames dancing in midair, and the BTH workers vanishing "even more rapidly."

But, said Whittle, after the soothing balm of several glasses of red wine taken in a local hotel, there was a simple explanation, and for a while he was confident that the combustion problem could be solved. But he was overoptimistic, and after test after test after test through that summer of 1937, all failures of one kind or another, a total redesign of the engine seemed essential. Yet, by now, there was almost no money, Whittle himself was in a near-hysterical mood, and the project seemed in dire danger of foundering. Moreover, the testing had become so dangerous that BTH insisted that any further experiments be conducted at a site seven safe miles away from its factory, in a disused foundry near the town of Lutterworth.

It was here that the project's fortunes turned. By now, the Air Ministry had decided to throw in a modest sum, largely because Henry Tizard had written glowingly of what he believed to be Whittle's genius, and Tizard was so widely respected that notice was taken at the highest levels of government. BTH also put in some funds, and new tests of Whittle's redesigned engine began in April 1938. The first run ended when a cleaning rag was sucked into the engine through the compressor fan. In May, a test run achieved a speed of 13,000 rpm, though it was brought to a costly sudden shutdown when nine of the turbine blades shattered, detached themselves from the disc, and blasted their way through the engine. It took four more months to rebuild it, and this time, instead of building just one combustion chamber, the engineers built ten of them, which enveloped the turbine rotor like insulating pillows and gave the engine a look of substance and heft and symmetry, ironically not too dissimilar from the radial piston engines the jet sought to supplant.

And this engine worked, finally. On June 30, 1939, less than ten weeks before the outbreak of World War II, an Air Ministry official came up to Lutterworth to inspect it, witnessed it running for twenty-eight minutes at a sustained speed of 16,000 rpm, and made a crucially important decision. Whittle's design was to be approved for the manufacture of a flight engine; and shortly thereafter, the Gloster Aircraft Company* was ordered to produce an experimental airplane that would be powered by it. The engine was to be designated the W1X; the plane, the Gloster E28/39.

It fell to the technical director at Gloster, a sober-sided pipe-smoking engineer named George Carter, to design the new craft. The ministry wanted it to be

---

* The firm had been established as the Gloucestershire Aircraft Company Limited in 1917, but changed its name to Gloster because many foreign customers found the original spelling too difficult to pronounce.

both a flying test-bed and a warplane, so it had to have four guns and be loaded with ammunition. But Carter said it should be small and light, weighing little more than a ton, and eventually won the government's agreement that he could leave the weaponry out of the first two prototypes. Building started in 1940, when war was fully under way and the Luftwaffe was bombing British cities with great enthusiasm, so Gloster, which had an all-too-visible factory and airfield near its home base, decided to move this highly secret project into an abandoned motorcar showroom, the Regent Garage, nearby, in the city of Cheltenham. A single armed policeman stood guard outside while, inside, a small band of craftsmen labored to finish the machine. No one, or no German, ever found out.

It is worth noting that during the run-up to the first British jet-powered flight, a German turbojet-powered aircraft had already been tested, on August 27, 1939, a week before the war's outbreak. The plane was the Heinkel He 178, and its engine was based on the design by Hans von Ohain back in 1933, mentioned previously. The German government was unimpressed by the craft, however, deriding it for being slow and having a combat endurance of only a few minutes. Berlin then eventually bowed to advice (offered to Hitler himself by the great German aircraft designer Willy

Messerschmitt) that jets used too much fuel. So the privately funded and developed Heinkel experiment, though technically the first jet-powered flight ever, proved an eventual failure.

The shrouds came off the British effort in the early spring of 1941, to reveal a sweet little aircraft, toy-like in its smooth and stubby simplicity, with a foot-wide, mouthlike air intake hole for a nose—and no propeller!—a jet pipe snugged in beneath the tail, a pair of wings, a sliding-door cockpit, and little else. The undercarriage was short and retractable—there was no need to have the plane high above the runway to prevent a spinning propeller from striking the ground. In short, the Gloster E28/39—the government's order number was 28; the production year was 1939—was simplicity itself, economical in look, in design, and in cost of materials.

It was completed some months before Whittle's engine, which still had myriad fine-tuning problems. At one stage, the entire engine was mounted in the tail assembly of a big Wellington bomber (air inlets replacing the gun turrets) to see how it would perform at altitude. It did well, and so, unbolted from the bomber, it was taken by truck to the Gloster test airfield near the Cotswold village of Brockworth, a town better known today for its annual midsummer cheese-rolling contest, when drunken locals try to pursue a huge round cheese as it is set thundering down a local hill. There, it was finally mated into George Carter's little aircraft: it sat directly behind the pilot, though with the fuel tank sandwiched between it and the pilot's back.

Unlike the cheese, the aircraft was kept firmly on the horizontal for its first trials, largely to see how it handled during taxiing. But the chosen test pilot, Gerry Sayer, was apparently unable to contain himself at the smoothness of the throttle controls, and at the rapid acceleration to full power of a near-vibrationless engine, and so took the plane off for a pair of hundred-foot hops along the runway, astonishing all, and prompting an American engineer standing on the wing of a Stirling bomber almost to fall onto the ground at seeing a propellerless aircraft roaring along a runway and lifting off, if only for a few seconds. He was told to disbelieve what he had seen. German agents might have been everywhere.

In the end, it was decided to take the aircraft (now semiofficially known as the Pioneer, which, though a name of slightly greater historical portent, never took off, as it were) up to the airfield at Cranwell, Whittle's old air force school. It was flatter (fewer cheese-rolling hills) and less populated, making it easier to keep the first flight secret.

This being Britain, it was the weather that now conspired to stall things, and the chosen day, May 15, 1941, dawned cloudy and cold. Whittle left for the engine assembly plant, where he had work to do on the next generation of engines the air force had selected to be put into what would be called Gloster Meteor fighters. He kept an eye on the skies, however, and when finally there were sufficient patches of blue sky "to mend a sailor's trousers," as the saying has it, he knew the evening would come clear. He drove back to Cranwell like the wind.

He was only just in time. As he suspected, Gerry Sayer already had the plane out on the long east–west runway. The breathless Whittle joined his colleagues from Power Jets and took a car to about the halfway mark, and waited there as they watched Sayer turning the doughty little craft into the bitterly cold westerly breeze.

Anticipating speed, Sayer secured his cockpit canopy. He set his trim to keep the nose slightly down, and retracted the flaps. Then he stood on the brakes and began to spool up the engine. When it was whining satisfactorily and the plane was bouncing against the brakes, he took his feet off the pedals. The craft bounded forward and began to accelerate toward the watery sun. It was 7:40 p.m. Dusk was falling. Whittle watched, clenching his fists with anxiety.

After about five hundred yards of steady acceleration, and with the engine now roaring lustily and flame
spearing from the after jet pipe, Sayer eased back the
stick. Effortlessly, with the aerofoils behaving in text-
book style and the engine never once faltering in its
delivery of a pure, whistling five hundred horses of
power, the tiny aircraft rose calmly and propellerless
into the evening sky. In seconds, the Pioneer was at a
thousand feet, and the watchers on the ground could
see as Sayer used the hydraulic accumulator to retract
the undercarriage. Suddenly, the plane, by now emitting a faint trail of dark smoke, looked like a smoothly
engineered bullet, vanishing into clouds that then closed
seamlessly behind her.

All that Whittle and his colleagues could then hear
was the even roar of the engine—a jet engine, the
first-ever confection of precisely engineered compressor blades and turbine blades and hot-sprayed fuel and
Newton's Third Law to rise and to fly in England, and
the first ever in the world to enjoy the support of a national government. For the coming minutes, there may
have been nothing to see but the clouds overhead, but
the timbre and volume and direction of the sound indicated to all below that, up there, Gerry Sayer was having
fun, was putting the little craft through her paces, was

being the exemplar of an old-school test pilot, and was inaugurating, officially, the Jet Age.

Then, after maybe a quarter of an hour, Whittle and his men heard the sound grow louder from the east, and then they could see her, glinting in the low sun and preparing to land. They saw the undercarriage come down; the flaps and spoilers were lowered, the speed reduced, the glide path achieved—until the plane was no more than ten feet above the rain-damp runway, moving so slowly and decorously it was almost hovering. At this point, Sayer cut the power way back, and the craft settled down for the final seconds of her first flight, then dropped gently onto the center line with the wheel supports bouncing under the weight, and then he turned his charge toward the waiting car and stopped, turning the throttle back to stop and silence the engine. All was quiet: there was no sound now except for faint residual radio chatter with the control tower, the creak of cooling metal on the fuselage—it was cold that night, and the engine parts had been very warm—the susurrus of blowing airfield grass as the breeze kicked up a little, and then, suddenly, the unmistakable sound of frantically running feet.

They were racing toward the plane. Frank Whittle, who thirteen years before had dreamed up the idea of an engine and had battled long and hard to get it made, and George Carter, who had designed the tiny craft that would be hoisted by it into the sky, and into the history books, ran without thinking across the taxiway, and together they reached up and grasped Gerry Sayer's hand and shook it with congratulations and expressions of relief. It was the spring of 1941. A new era had begun.

Yet there was no Ministry of Information film crew to take note, no journalist on hand, no one from the BBC, no photographer, save for one amateur who took a blurry picture of a grinning Whittle reaching up to the lip of the cockpit and offering his congratulations and thanks.

It was not until the New Year of 1944, fully two years and eight months later, that the British public was told of the new invention, of the new age that had stolen up on them, unbeknownst to almost all. "Jet Propelled Aeroplane," said the *Times*, on page 4. "Success of British Invention": "After years of experiments Britain now has flying a fighter aeroplane propelled by a revolutionary type of power unit, the perfection of which represents one of the greatest steps forward in the history of aviation. The new system, known as jet propulsion, does away with the need for an orthodox engine and also for an air-screw."

Frank Whittle's name is mentioned four paragraphs in, as is the fact that the U.S. government was apprised of the success of the first flight within weeks of its occurrence, in July 1941. Yet the British public, which had footed the bill, was kept firmly in the dark. Likewise the American public, who were told the news of the new engine on the same day: January 6, 1944.

Frank Whittle, initially honored—King George VI conferred a knighthood—and somewhat revered, did not have as happy a time in postwar England as one might think he deserved. Power Jets was nationalized, and its chief engineer sidelined, put out to pasture. He traveled, he lectured, he wrote, and he particularly savored his election to a fellowship of the Royal Society. He won prizes, the most valuable of which, at around half a million dollars, he decided generously to split with Hans von Ohain, the German inventor whose Heinkel-powered plane had been the true first to fly with a turbojet engine. Whittle argued often for the good sense of building a supersonic passenger plane, and badgered officials long before Concorde was a drawing board dream. But no one listened, and by 1976, with his marriage failed, he decided to light out for America, and spent his final years in a suburb of Washington, DC.

Occasionally he was called back home. He returned to be presented with the Order of Merit by Queen

Elizabeth in 1986; and again when there was a bit of a fuss around the fiftieth anniversary of his former engine company's creation, in 1987; and then, with his son Ian Whittle piloting, he came to London and flew happily on a Cathay Pacific 747 passenger aircraft nonstop to Hong Kong.

It was in one small and curious way a memorable flight. For, back then, when Kai Tak Airport was the only commercial airfield in the then–British colony, most inbound flights had to make an alarming last-minute course change in order to land safely. Standing instructions for the approach required that the plane come into the colony's airspace from the west and, losing height rapidly, head directly toward an enormous red-and-white checkerboard that had been obligingly painted onto the rockface of a mountainside. When the plane was just a mile distant, less than twenty seconds from closing hard with the rocks, the pilot had to make a sudden sharp, thirty-seven-and-a-half-degree turn to starboard, a maneuver that, if faultlessly performed, then allows for a direct low-altitude approach onto Kai Tak's runway 013.

Anyone not warned beforehand about this maneuver can be severely alarmed—and Frank Whittle, who had been sitting calmly in the cockpit behind his son during the cruise and was now preparing for a routine landing, was indeed somewhat bothered by what, for a few seconds, seemed an inevitable crash. But the required maneuver, invariably perfectly timed and precisely accomplished by pilots of long experience with this most exotic of eastern approaches (his son this day included), put the aircraft down a few moments later, and with customary exactitude.

The plane that day had been powered by four Rolls-Royce jet engines,* all of which had fired perfectly to complete this dramatic maneuver. It was also a Rolls-Royce jet engine, but a very much more powerful variant, and built for a very much larger aircraft, that, almost a quarter of a century later, failed so dramatically over Indonesia. The official postmortems, published three years later in Australia, went some way

---

* Rolls-Royce started making aircraft engines in 1915, little more than a decade after launching its first motorcar. The firm branched into jet engines in 1946, and in the early 1950s, the Avon was used to power the Canberra bomber for the Royal Air Force and the ill-fated Comet for British Overseas Airways Corporation, BOAC. Despite many stumbles, including bankruptcy and nationalization (later reversed), the company, making aero engines for more than a century, remains a formidable player in the jet engine market, having made some fifty thousand thus far.

toward illuminating the formidable technical problems and challenges involved in the making of a modern high-power, high-performance jet engine.

Although a modern jet engine is, upon close inspection, a thing of the most fantastic complexity, it is easy to believe this is not so. Its exterior cowling is so clean and smooth; the fan blades at its open mouth turn with such slow elegance; the sounds it emits, even at full throttle, have such a sonorous harmony about them, that it is tempting to imagine all is the purest simplicity within. In fact, once the covers are removed, everything inside is a diabolic labyrinth, a maze of fans and pipes and rotors and discs and tubes and sensors and a Turk's head of wires of such confusion that it doesn't seem possible that any metal thing inside it could possibly even move without striking and cutting and dismembering all the other metal things that are crammed together in such dangerously interfering proximity. Yet work and move a jet engine most certainly does, with every bit of it impressively engineered to do so, time and again, and under the harshest and fiercest of working conditions. And nowhere is the environment more harsh or more fierce than in the high-pressure section of the turbine, the fattest, smoothest, and, to the outsider, most innocent-looking part of a jet engine, with nothing (such as a fan) to be seen moving and nothing (such as a hot exhaust blast) to be felt or heard to any degree.

There are scores of blades of various sizes in a modern jet engine, whirling this way and that and performing various tasks that help push the hundreds of tons of airplane up and through the sky. But the blades of the high-pressure turbines represent the singularly truest marvel of engineering achievement—and this is primarily because the blades themselves, rotating at incredible speeds and each one of them generating during its maximum operation as much power as a Formula One racing car, operate in a stream of gases that are far hotter than the melting point of the metal from which the blades were made. What stopped these blades from melting? What kept them from disintegrating, from destroying the engine and all who were kept aloft by its power? It seems at first blush so ludicrously counter-intuitive: that a piece of normally hard metal can continue to work at a temperature in which the basic laws of physics demand that it become soft, melt, and turn to liquid. How to avoid such a thing is central to the successful operation of a modern jet engine.

For, very basically, it turns out to be possible to cool the blades by performing on them mechanical work of a quite astonishing degree of precision, work which allows them to survive their torture for as many hours as

Five conjoined high-pressure turbine blades in a jet engine, fashioned from single-crystal titanium alloy and peppered with tiny holes that allow a rush of cool air to prevent them from melting in the chamber's lethally hot atmosphere.

the plane is in the air and the engine is operating at full throttle. The mechanical work involves, on one level, the drilling of hundreds of tiny holes in each blade, and of making inside each blade a network of tiny cooling tunnels, all of them manufactured at a size and to such minuscule tolerances as were quite unthinkable only a few years ago.

Inevitably, it was commerce that provided the spur for all this work—although the jet engine makers who worked secretly for "the dark side," creating technologies for bombers and stealth fighters and their like, made as-yet-unacknowledged contributions, too, and about which plane makers still cannot talk. The start of work on turbine blade efficiency began in the 1950s, just as soon as piston-engined aircraft began to be eased out of the world's main skyways, and as soon as jet engines, initially developed for military use, were being redesigned in ways that made economic sense for hauling passengers and freight over long distances at high speed. Aircraft such as the Viscount, the Comet, the Tupolev Tu-104, the Convair 880, the Caravelle, the Douglas DC-8, and, from 1958, the best known of all narrow-bodied jets, the Boeing 707, began to sweep the field. The engines with which they were equipped (the De Havilland Ghost; Pratt and Whitney's JT3C and JT3D; Rolls-Royce's Avon, Spey, and Conway; and for the two hundred Tupolev Tu-104s that Moscow built, the little-known Mikulin AM-3 turbojet) were all of their time state-of-the-art high-precision machines.

By today's standards, these older engines were relatively primitive, being noisy, fuel-hungry, underpowered, and inefficient. Yet all this started to change, once again in the 1970s, as more and more aircraft were needed to fly over greater and greater distances and at higher and higher speeds. To produce the necessary thrust for the big and more economical wide-bodied jetliners that growing numbers of passengers and hard-pressed airline accountants alike were demanding, and to produce that thrust quietly and efficiently and with something of a nod to the growing environmental concerns of the latter half of the century, the new jet engines had to be huge and astonishingly powerful. They had to compress their inswept air (as much as one ton of it sucked in every second) to unimaginable pressures, they had to burn their fuel at unimaginable temperatures, and they had to create an interior holocaust, a maelstrom of fire, that tested every molecule of every metal piece that whirled and careened around inside.

This is where Rolls-Royce's internal Blade Cooling Research Group, founded in the early 1970s, plays its part in the saga. The group's mission was simple enough: solve the problem of keeping those high-pressure turbine blades from melting, and then jet engines could be made that would give out all the power anyone might need. For the axiom of turbinology is a simple one: the hotter the engine is run, the greater the spare pressure, and the higher the jet velocity. The hotter, in other words, the faster.

At the same time, though, the hotter the engine environment, the bigger the problem for the turbine blades. For while one might suppose the first task of a turbine blade is to drive the engine's compressor, it actually is not. That is its secondary task. Its first task is quite simply to survive.

In Whittle's engines, and the military jets that were built immediately after his invention turned out to be a proven success (and in the civilian world's turbojet engine of the Vickers Viscount and the pure jet engines of the Comet, which was to become the first-ever commercial jetliner), the survival of the turbine blades was not a major issue.

They were critically important components, of course. The first blades that Whittle made were of steel, which somewhat limited the performance of his early prototypes, since steel loses its structural integrity at temperatures higher than about 500 degrees Celsius. But alloys were soon found that made matters much easier, after which blades were constructed from these new metal compounds in ways that met most of the challenges of the earliest engines. They were shaped to meet and extract energy from the peculiar violent vortices of the hot gases that swirled about them. They were fixed to the disc that carried them in a way that could manage the otherwise intolerable stresses of being whirled around at hundreds of revolutions each minute. Their shape was such that they managed to extract with remarkable efficiency the power from the chemical reaction between the hot compressed air and the fuel (gasoline in Whittle's first laboratory, kerosene later on) delivered to them. They did not run the risk of melting, though, because the temperatures at which they operated were on the order of a thousand degrees, and the special nickel-and-chromium alloy from which they were made, known as Nimonic, remained solid and secure and stiff up to 1,400 degrees Celsius. There was adequate leeway between the temperature of the gas and the melting point of the blades. That would change, though, in the 1960s and '70s. The leeway steadily diminished, and soon it finally vanished altogether.

For, by then, the demands made on the next generation of engines required that the gas mixture roaring out from the combustion chamber be heated to around 1,600 degrees Celsius, and even the finest of the alloys then used melted at around 1,455 degrees Celsius. The metals tended to lose their strength and become soft and vulnerable to all kinds of shape changes and expansions at even lower temperatures. In fact, extended thermal pummeling of the blades at anything above 1,300 degrees Celsius was regarded by early researchers as just too difficult and risky—unless someone could come up with a means of keeping the blades cool.

A team of about a dozen Rolls-Royce engineers promptly did just that. They worked out that it should be possible, with highly precise machining and the mathematical abilities of very powerful computers, to create an ultrathin film of relatively cold air that would swaddle each blade as it whirled around inside the engine, and which would protect it, thermally, from the hellish atmosphere beyond. The layer of cold air need be less than a millimeter thick, but if it managed to maintain its own integrity as the blade spun around, then the swaddled blade would also.

But where to acquire the cold air, inside a jet engine? The source was hidden, it turns out, in plain sight. After much pondering and experimenting, it was realized that the cooler air could come directly from the huge tonnage of atmosphere being sucked in by the fan at the engine's front. Most of that air bypasses the engine (for reasons that are beyond the scope of this chapter), but a substantial portion of it is sent through a witheringly complex maze of blades, some whirling, some bolted and static, that make up the front and relatively cool end of a jet engine and that compress the air, by as much as fifty times. The one ton of air taken each second by the fan, and which would in normal circumstances entirely fill the space equivalent of a squash court, is squeezed to a point where it could fit into a decent-size suitcase. It is dense, and it is hot, and it is ready for high drama.

For very nearly all this compressed air is directed

straight into the combustion chamber, where it mixes with sprayed kerosene, is ignited by an array of electronic matches, as it were, and explodes directly into the whirling wheel of turbine blades. These blades (more than ninety of them in a modern jet engine, and attached to the outer edge of a disc rotating at great speed) are the first port of call for the air before it passes through the rest of the turbine and, joining the bypassed cool air from the fan, gushes wildly out of the rear of the engine and pushes the plane forward.

"Nearly all" is the key. Some of this cool air, the Rolls-Royce engineers realized, could actually be diverted *before* it reached the combustion chamber, and could be fed into tubes in the disc onto which the blades were bolted. From there it could be directed into a branching network of channels or tunnels that had been machined into the interior of the blade itself. And now that the blade was filled with cool air—cool only by comparison; the simple act of compressing it made it quite hot, about 650 degrees Celsius, but still cooler by a thousand degrees than the post–combustion chamber fuel-air mixture. To make use of this cool air, scores of unimaginably tiny holes were then drilled into the blade surface, drilled with great precision and delicacy and in configurations that had been dictated by the computers, and drilled down through the blade alloy until each one of them reached just into the cool-air-filled tunnels—thus immediately allowing the cool air within to escape or seep or flow or thrust outward, and onto the gleaming hot surface of the blade.

If the mathematics is performed correctly—and it is here that the awesome computational power that has been available since the late 1960s comes into its own, becomes so crucially useful—and if the placing of all these pepperings of minute holes is correctly achieved, with some holes on the blade's leading edge, some on its chubby little body, some along the trailing edge, then this cool air will form an unimaginably thin film of comforting relative frigidity, wrapping itself around the blade and coating its whirling surface like a silvery insulating jacket. It is this, then, that allows the blade to survive the blistering heat of the onrushing fuel-air mixture, which the combustors have just set alight.*

---

* The first Rolls-Royce engine to employ this kind of blade cooling, in the late 1960s, was the RB211, which proved to be so costly to develop that it brought the company to insolvency, and to its nationalization for seven years by the British government. One of the early problems came about from the use of carbon fiber blades for the main external fan. By regulation, these had to be tested for resistance to bird strikes. A cannon fired a five-pound chicken at the spinning blades, which, to universal dismay, promptly shattered into thousands of pieces. These blades were replaced eventually with compressor blades made of

All who see such a jet engine turbine blade, and who know something of its manufacture, see in its making the most sublime of engineering poetry, much like the finest of Rolls-Royce motorcars, one might say—the Silver Ghosts of eighty years before had many of the qualities of perfection that are engineered into today's better aircraft engines. Each of the Rolls-Royce nickel alloy blades (which weigh less than a pound, are mostly hollow but sensationally strong, can fit easily into the palm of the hand, and, as it happens, are also, for now, essentially made by hand) is cast in a top-secret factory near Rotherham, in northern England. The most proprietary and commercially sensitive aspect of the blades, aside from the complex geometry of the hundreds of tiny pinholes, is the fact that the blades are grown from, incredibly, a single crystal of metallic nickel alloy. This makes them extremely strong—which they need to be, as in their high-temperature whirlings, they are sub-

---

titanium, but this took time and money and, for a while, cost the company its livelihood.

However, the RB211 did eventually outperform its main American competitor, the Pratt and Whitney JT9D, used in early jumbo jets. NASA statistics show that, in the 1970s, the JT9D had an average of one engine shutdown per transatlantic flight, whereas the RB211 had one shutdown every ten crossings. It was fortunate the aircraft had four engines, and that the passengers never came to know.

jected to centrifugal forces equivalent to the weight of a double-decker London bus, of around eighteen tons.

There is a delicious irony here, however. For although, as one might expect, to make such a blade requires techniques displaying the very highest order of precision and computational power, they are combined with another means of manufacturing that is of the greatest antiquity. The "lost-wax method" was known to the Ancient Greeks, for whom precision was a wholly unfamiliar concept.* It is employed specifically in this case to allow the creation of the cooling tunnels within the blade; and the wax is melted out, as in Athenian days, just before the molten alloy is poured into the ceramic mold, which is now, absent the wax, busy with the network of voids for the eventual cooling air.

Creation of the blade's single-crystal structure is encouraged at this very point in the long and cumbersome manufacturing process, and is the company's most closely guarded secret. Very basically, the molten metal (an alloy of nickel, aluminum, chromium, tantalum, titanium, and five other rare-earth elements

---

* Except, perhaps, to the makers of the Antikythera mechanism, mentioned in chapter 1, which, though it has the look of a high-precision instrument, manages to be wholly lacking in accuracy. It was made, however, more than two thousand years ago, so its makers can perhaps be forgiven.

that Rolls-Royce coyly refuses to discuss) is poured into a mold that has at its base a little and curiously three-turned twisted tube, which resembles nothing more than the tail of P. G. Wodehouse's Empress of Blandings, the fictional Lord Emsworth's prize pig. This "pigtail" is attached to a plate that is cooled with water, and the whole arrangement, once it is filled with liquid metal, is slowly withdrawn from the furnace, allowing the metal, equally slowly, to solidify.

This it does, first, at the cool end of the pigtail, but because the mold here is so twisted, only the fastest-growing crystals and those with their molecules distributed with what is called a face-centered cubic arrangement, for complex reasons known only to students of the arcana of metallurgy, manage to get through. And through this magic of metallurgy, the entire blade then assembles itself from the one crystal that makes it along the pigtail, and ends up with all its molecules lined up evenly. It has become, in other words, a single crystal of metal, and thus, its eventual resistance to all the physical problems that normally plague metal pieces like this is mightily enhanced. It is very much stronger—which it needs to be, considering the enormous centrifugal forces that dominate its working life.

Having now created the single-crystal blade, it remains only to dissolve away any of the substances that remain in its core, and then to use a technique called electrical discharge machining to drill the hundreds of tiny holes down into the cooling channels. Electrical discharge machining, or EDM, as it is more generally known, employs just a wire and a spark, both of them tiny, the whole process directed by computer and inspected by humans, using powerful microscopes, as it is happening. The process is all but silent, and it is in many ways more melting than drilling.

Here, however, comes an important moment in the story, one that has crept into the narrative all too stealthily.

The making of high-pressure turbine blades has long required the absolute concentration of legions

Despite the apparent fantastical complexity of a modern jet engine—here a Rolls-Royce Trent, four of which power the enormous Airbus A380 double-decker superjumbo jets—there is essentially only one moving part, a rotor that passes through the entire length of the engine, from fan to exhaust.

of workers, men and women with decades of experience in hand-eye coordination and a studiously learned degree of extreme manual dexterity. These "blade runners," as it were, have for years past learned to manage, for instance, the complexities and eccentricities of the cooling-hole drilling machines—and the more complex the engines, the more holes need to be drilled into the various surfaces of a single blade: in a Trent XWB engine, there are some six hundred, arranged in bewildering geometries to ensure that the blade remains stiff, solid, and as cool as possible.

Yet human lives, those of the aircraft passengers and crew, are dependent on the engine's not self-destructing in flight. The vanishingly small number of occurrences of this kind of incident is based to a large degree on the integrity of these human-made engine blades. Because there is no doubt of the blades' immense importance, it is worth noting that their integrity owes much to the geometry of the cooling holes that are being drilled, which is measured and computed and checked by skilled human beings. No tolerance whatsoever can be accorded to any errors that might creep into the manufacturing process, for a failure in this part of a jet engine can turn into a swiftly accelerating disaster.

This stark realization, that lives depend on the perfection of these blades, brings this one industry to a critical moment, a crucial development—the first in the story, perhaps, and one that would be unimaginable to precision's originators, to engineers such as John Wilkinson or Joseph Bramah, Henry Maudslay or Joseph Whitworth, or indeed, to Henry Royce himself. Engineering, in this one field to start with, seems now to have reached a degree of sophistication in which the rigorous demands of modern precision have come for the first time to outstrip the abilities of humankind to meet them.

Up until this point, the processes—whether it is the making of a cylinder or a lock or a gun or a car; the boring or the milling or the grinding or the filing; the directing of the lathe or the tightening of a screw or the measuring of flatness or circularity or smoothness—invariably involve some kind of human agency. Yet now, in this one field to begin with, but in many more as the tolerances shrink still further and limits are set to which even the most well-honed human skills cannot be matched, automation has to take over. The Advanced Blade Casting Facility can perform all these tasks (from the injection of the losable wax to the growing of single-crystal alloys to the drilling of the cooling holes) with the employment of no more than a handful of skilled men and women. It can turn out one hundred thousand blades a year, all free of errors—or as far as anyone knows.

Once, the most troubling consequence of the introduction of precision machinery was the displacement of unneeded workers, who were understandably vexed. Nowadays, it is perhaps the relative paucity of human supervision in engineering fields where human lives are at stake that has steadily become a more pressing concern.

"Our people are fantastically skilled," remarked the manager of manufacturing at the new plant, "but they're human, and no human is going to produce the same quality of work at the end of a shift as they do at the beginning." Precision engineering, in this industry in particular, does now appear to have reached some kind of limit, where the presence of humans, once essential to maintaining the attainment of the precise, can on occasion be more of a drawback than a boon—as the investigation into the Qantas Airways jet engine failure amply illustrates.

In the immediate aftermath of the incident, the airline grounded all six of the Airbus A380s in its fleet, and angrily threatened lawsuits against Rolls-Royce because of the commercial impact of the accident. Yet anger plays no part in an investigation into an aircraft accident, and the Australian government's Transport Safety Bureau then took the lead in determining what or who was at fault. The official report, issued almost three years after the accident, in June 2013, turned out to be a damning indictment of an industrial culture that had taken for granted the need for absolute precision to be applied consistently in the making of every single part of a modern high-performance jet.

For it transpired that the fate of this engine, of this plane, of these passengers and crew, of the reputation of an airline and of the engine maker, all turned on the performance of one tiny metal pipe. It was a pipe no more than five centimeters long and three-quarters of a centimeter in diameter, into which someone at a factory in the northern English Midlands had bored a tiny hole, but had mistakenly bored it fractionally out of true.

The engine part in question is called an oil feed stub pipe, and though there are many small steel tubes wandering snakelike through any engine, this particular one, a slightly wider stub at the end of longer but narrower snakelike pipe, was positioned in the red-hot air chamber between the high- and intermediate-pressure turbine discs. It was designed to send oil down to the bearings on the rotor that carried the fast-spinning disc. It needed to have a filter fitted into it, so the stub end of it had to be reamed out to make certain it could accommodate the metal ring of this filter.

The tube and the assembly around it had been

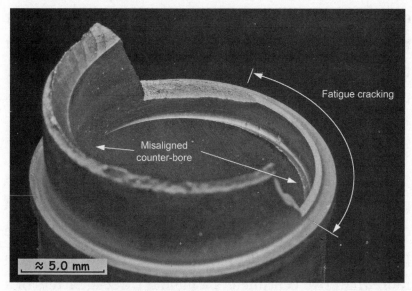

Fatigue cracking

Misaligned counter-bore

≈ 5.0 mm

The fractured oil feed stub pipe that failed due to metal fatigue caused by slightly misaligned machining that left the pipe a little thinner on one side. Fatigue cracking probably began upon takeoff in Los Angeles, and worsened after takeoff from London. A minute after takeoff from Singapore, the pipe broke open and spewed hot oil over the rapidly spinning rotor.

manufactured in a Rolls-Royce plant, Hucknall Casings and Structures, sometime in the spring of 2009. In normal circumstances, it would have been trivially easy to machine the pipe for the filter fitting, and to do so to the exacting standards laid down by the designers of the engine. But for this particularly complicated part of this particular engine, it was decided to complete first the entirety of the hub assembly that separates the high-pressure from the intermediate-pressure areas of the engine—and then and only then, once the pipe had been fitted into place in this assembly, to drill out the tube to its design specifications. This proved to be exceptionally difficult, however, because now parts of the pipe could not be readily seen, as other parts of the fully assembled hub and newly made welds of its various pieces obscured the engineers' clear sight of it.

These engineers did the best they could, but in the end, the tiny pipe that would eventually go into the turbine of the engine that would be suspended from the portside wing of the Qantas A380 was machined improperly: the drill bit that did the work was misaligned, with the result that along one small portion of its circumference, the tube was about half a millimeter too thin.

The assumption is that, during manufacture, the hub assembly somehow moved a tiny amount as it was being drilled, with the result that the drill bit moved fractionally closer to one wall of the pipe, reducing it to what would be dangerously vulnerable thinness. More dangerously still, the quality-control departments at Hucknall Casings, and the computer-driven machines that check the conformance of all critical parts of an aircraft, passed the tube as being satisfactory. The badly made part should have thrown up all manner of red flags. It should have been discarded—a high-pressure turbine blade, deemed to be an absolutely critical and safety-critical part of an engine, would have been tossed out and smashed for an error far less significant than the error in this tube.

Yet, for reasons that have much to do with what is euphemistically called the "culture" of that particular facility within Rolls-Royce's immense engineering establishment, the stub pipe passed all its inspections. A potentially weakened engine component made its way all along the supply chain until it was placed into the engine, and there to await its inevitable breakage—and the equally inevitable destruction of the entire engine. It should have failed inspection, but it didn't. It just failed in real life.

Metal fatigue is what did it. The aircraft had spent 8,500 hours aloft, and had performed 1,800 takeoff and landing cycles. It is these last that punish the mechanical parts of a plane: the landing gear, the flaps, the brakes, and the internal components of the jet engines. For, every time there is a truly fast or steep takeoff, or every time there is a hard landing, these parts are put under stress that is momentarily greater than the running stresses of temperature and pressure for which the innards of a jet engine are notorious.

And from what can be divined, the weakness in the wall of the stub pipe gradually transformed itself into a fatigue crack. That crack first opened, very slightly, investigators believe, when the plane took off from a short runway in Los Angeles two days prior to its take-off from Singapore. The fissure then began to spread and to open a little more when the plane landed in London; it came under still further stress when the plane took off from Heathrow bound for Singapore, and once again when it landed at Changi Airport hours before its departure for Sydney.

Ninety seconds after this midmorning takeoff, with the plane climbing steeply and the engine running at 86 percent of full power, throwing out more than 65,000 pounds of thrust, the crack finally opened fully and the pipe ruptured. A spray of hot oil immediately hissed out into the void between the high- and intermediate-pressure turbines, where the temperature was already at some 400 degrees Celsius. The auto-ignition temperature of the oil was 286 degrees, and the jet of oil mist became like a high-powered flamethrower, pouring fire against the huge, heavy, rapidly spinning turbine disc.

After a few seconds of such ferocious heating, the disc expanded, lost its shape, began to wobble furiously, and then eventually broke, and hurled its fractured segments at hundreds of miles an hour out through the engine and out of the casing and, in two instances, through the left wing of the plane and, in a third, clear through the bottom of the fuselage. A brief flash fire broke out inside the left wing, but mercifully did not spread; the damage to the hydraulics and electrics occasioned by the shrapnel caused a series of major failures to the aircraft systems. All ended well, thanks in large part to the crew, as the Australian government report noted.

What the report also noted, however, were the failures within Rolls-Royce: the failure to machine a critical part properly, the failure to keep proper records, the failure to inspect properly, and the failure to reject what were called "non-conforming" parts, and to allow them to pass into service, with potentially lethal consequences. The delivery of such engines to Qantas was far from unique: hasty inspection soon after the accident revealed that scores of Hucknall-made oil pipes with misaligned walls thinner than half a millimeter had already gone into service, with the consequence that no fewer than forty engines, in use by Singapore Airlines and Lufthansa, and on all the remaining five Qantas aircraft, needed to be withdrawn from use and repaired.

It was a costly mistake by Rolls-Royce, for not only were there internal repercussions, expensive repairs, staff changes, procedural reforms, and public relations nightmares, but also Qantas was paid some eighty million dollars in compensation; the Rolls-Royce balance sheet for the year after the event showed a net loss to the company of seventy million dollars. The firm insists that such an error is unlikely to happen again, and that all necessary precautions have been taken, at Hucknall and elsewhere.

Buried deep within the Australian government report on the accident is one paragraph that seems especially relevant to the wider issues posed by the ever-increasing precision of modern machinery. Like much of the 284-page report, the paragraph is rich with jargon, but the basic message shines through nonetheless:

> *Large aerospace organizations are complex socio-technical systems made up of organized humans producing highly technical artefacts for complex systems, such as modern aircraft. Due to the inherent nature of these complex sociotechnical systems, their natural tendency is to regress if not constantly monitored—and occasionally even when monitored vigorously. This natural regression can occur due to the pressure applied via global economic forces, the requirement for developing growth, profit and market share . . .*

"Highly technical artefacts for complex systems"—shorthand, or rather, bureaucratic longhand, for *ultraprecise machines*, such as the Trent 900-series jet engine. Maybe, this incident will suggest to some, modern machines of certain specific kinds are being made now with just too much precision, with too much complexity, for it to be prudent for humans to participate in the making of them. If this is true, it might reasonably prompt the question: could we be seeing here the beginning of the upper limits of our ability to manage the kind of precision we think we need?

Or maybe precision is itself reaching some kind of limits, where dimensions can be neither made nor measured—not so much because humans are too limited in their faculties to do so but, rather, because as engineering reaches ever downward, the inherent properties of matter start to become impossibly ambiguous. The German theoretical physicist Werner Heisenberg, in helping in the 1920s to father the concepts of quantum mechanics, made discoveries and presented calculations that first suggested this might be true: that in dealing with the tiniest of particles, the tiniest of tolerances, the normal rules of precise measurement simply cease to apply. At near- and subatomic levels, solidity becomes merely a chimera; matter comes packaged as either waves or particles that are by themselves both indistinguishable and immeasurable and, even to the greatest talents, only vaguely comprehensible.★

In the making of the smallest parts for today's great jet engines, we are reaching down nowhere near the limits that so exercise the minds of quantum mechanicians. Yet we have reached a point in the story where we begin to notice our own possible limitations and, by extension and extrapolation, also the possible end point of our search for perfection. There may be an event horizon coming into view—and if so, then the work being performed at the jet engine makers of the world, where so much depends on such a testing level of precision, acts as a signpost, a warning that an end of sorts may be in sight.

Perhaps this sense of technical foreboding is true—as far as the making of machines and devices that are directly applicable to human-scale activities is concerned. Then again, to go beyond this, to venture into other worlds and to deal with other universes, maybe the lim-

---

★ Even Richard Feynman, everyone's favorite twentieth-century polymath, who won the 1965 Nobel Prize for Physics, famously asserted, "I think I can safely say that nobody understands quantum mechanics."

its that seem about to challenge human competence can in fact be pushed even higher and higher. Maybe, in these other worlds, precision can be further and further refined, with no end to its limits in sight.

Out in space, for example, all may be very different.

Chapter 7

(Tolerance: 0.000 000 000 000 1)

Through a Glass, Distinctly

The fate of human civilization will depend on whether the rockets of the future carry the astronomer's telescope or a hydrogen bomb.

—SIR BERNARD LOVELL, *The Individual and the Universe* (1959)

Murder most foul was committed one quiet summer's evening in a leafy South London park, but no one happened to notice it—not until the moment a fashion photographer, working quietly in his darkroom, enlarged and enlarged an otherwise innocent black-and-white image that he had taken in the park a while before and saw, or thought he saw, hidden in the trees, a hand with a gun and a body in the grass.

His film stock was grainy, and the enlarged pictures were blurred, but the images, all part of the story line from Michelangelo Antonioni's Oscar-nominated movie, *Blow-Up*, remain to haunt us to this day, and though the movie was about very many more things than murder, they serve as a reminder of the unassailable power of the camera to render random moments, sometimes quite inadvertently, into permanent historical truth—as I have lately come to know.

I work in an old timber-framed barn, a onetime granary built in upstate New York back in the 1820s. It was a tumbledown ruin when I bought it, and so I had its posts and beams trucked down to where I live, in a remote hamlet in the hills of western Massachusetts, and saw it rebuilt there in the summer of 2002. The internal arrangements of this modest little building allow for someone to look down from an upper gallery onto the confused mess that is my desk, fifteen feet below.

Because the barn is quite old, and because the phenomenon of breathing new life into old and decaying farm structures and renewing them as a living part of today's New England scenery was thought interesting, a photographer turned up one afternoon. He said he was working on a book on barn restorations, and once I happily gave him free rein, he spent some hours taking pictures, including some, from the gallery, of my paper-strewn desk below.

The images duly appeared in a rather handsome coffee-table book about the barn-rebuilding phenomenon. As a courtesy, I was duly sent a copy. I spent an evening admiring it (though, in truth, mostly envying barns of far greater grandeur than my own modest granary-that-was) before filing the volume away on the shelves and thinking no more of it.

Except, it turned out that someone quite unknown to me bought a copy of the book, too, and professed a liking for the little study structure he came across on page 61. Whether he was a fan of *Blow-Up*, I never knew, but he thought he might be able to find out just who it was who lived and worked there.

For on the desk in the picture was a copy of the *New York Review of Books*, half-covered with other litter: magazines and books and papers. The purchaser of the barn book espied that at the lower-right-hand side of the *Review* was an address label, small and barely noticeable to most. But to this fellow, it provided a possible source of information—if, that is, the lens that took the photograph was good enough for the label to be read when greatly magnified.

So he cut off the front cover of the *Review*, separating it from the other mess on the desk, and subjected it to ever-increasing degrees of magnification. The small and indistinct letters duly become ever larger and larger—until, even though the pixels of the printed image made for some eventual confusion, after four or five iterations of expansion, my name and address became legible. And all of a sudden, this mystery man knew who it was, most probably, who owned or lived in or made use of the barn. He got in touch.

And though the process sounds at this remove somewhat Peeping Tom–like, even faintly sinister, it turned out not to be so at all—the inquirer was entirely pleasant and most interesting; determined, slightly obsessive, maybe a trifle "on the spectrum," as is said today. He was a retired vascular neurosurgeon. He was a keen photographer. He was endlessly, preternaturally curious—polymathic, one might say—and he was fascinated most especially with the capabilities of precise optics in allowing for forensic detection, and with all the intellectual satisfaction that this could bring him.

As for most English schoolboys—for most schoolchildren everywhere, I daresay—lenses played a not insignificant part in my life. My first (most of which back in the 1940s were made of glass, plastic in those days being hardly good enough, and polycarbonates almost unknown) were all double-sided convex magnifiers. The first such lenses were used for trivia and for mischief: for examining tadpoles and peering at insufficiently detailed pictures in naturist magazines, for starting campfires, or for waking other boys foolishly unwary enough to fall asleep in the sunshine—a brief focus of sunlight on a bare arm would bring the deepest sleeper fully awake in seconds.

Better-quality lenses became more important to me when I was about ten and I became fascinated by phasmids, or stick insects. I would breed them—their homes were my mother's old Kilner jars filled with privet leaves from our garden hedge—and sell them to my classmates, threepence a time. But stick insects often develop strange microscopic problems—they sometimes find themselves unable to shake from their feet (of which, being insects, they have six) the egg cases from which they are born. Microsurgery, involving a needle, a fine tweezer, and my trusty times-ten magnifier, usually did the trick.

Then came gathering maturity. I went on to collect stamps, and amassed a collection of several magnifiers: a square-shaped lens to view the smaller stamps in full, a jeweler's loupe that I screwed into my eye and used for counting perforations and for spotting franking mishaps, and a heavy glass implement that looked like a paperweight but that, when I swept it across an album page, would let me display my collection, duly enlarged, to any curious passerby.

Precision optics (which generally meant expensive optics, and consequent pleas to the parents for funds) became of interest only when I hit fourteen or so, and needed, as I saw it, a microscope. Money was always short, but by rooting through secondhand shops and street barrows, I eventually acquired a range of those, too (made by firms such as Negretti and Zambra, Bausch and Lomb, Carl Zeiss), all in handsome wooden cases with slots for the changeable eyepieces and smaller slots for the magnifying lenses. I recall that there was a 1950s version of today's pixel envy, which had youngsters arguing over whose instrument offered the highest magnification. Given that we were looking at samples of pond water to spy out examples of *Daphnia*, or seawater to find those little pointed slivers of *Amphioxus*, and had neither the knowledge nor the equipment to probe much further into the world that Galileo and van Leeuwenhoek had bequeathed to us, there was little value in going beyond three-hundred-times magnification. I rather think some of my lenses allowed magnifications of a thousand, which was useless to my clumsy hands, which would knock something out of the field of view in an instant at what seemed like rocket speed. Some adolescent members of the school microscope club claimed to have seen their own spermatozoa, which struck me back then as both doubtful and disgusting, and also requiring an improbably handsome degree of magnification.

And then I bought a camera. A Brownie 127, first of all, with its plastic Dakon lens—a fixed aperture of f/14,★ a focal length of 65 mm, and a fixed shutter speed of a fiftieth of a second. I would take the rolls of exposed film to a small drugstore in the Dorset market town of my boarding school, and the chemist there who developed and enlarged the black-and-white images would encourage me, thinking my work had some small merit—or else, more probably, trying to get me to buy some of his small selection of cameras. I eventually caved in to his flattery and bought a 35 mm

---

★ The f number of a lens, seldom explained, is essentially a measure of how much light can get from the outside world into the inside of a camera. The number is very simply calculated by dividing the focal length of the lens (i.e., the distance from the center of the lens to where it focuses light on the film, or the sensor, in the back of the camera) by the diameter of the opening in the lens. A Brownie 127 lens with a focal length of 65 mm would need to have, in order to have its f number 14, a fixed aperture about 4 mm across.

Voigtländer* camera from him, a decision that sent me on a road that progressed over the years through a long trail of cameras that all used 35 mm film, most of them initially made in Japan by companies such as Pentax, Minolta, Yashica, Olympus, Sony, Nikon, and Canon.

Finally, one day in 1989 in Hong Kong, where I was living at the time, a young Cantonese salesman persuaded me that what I really needed was a quiet, compact, reliable, super-precise, and very sturdy 35 mm film camera that would be suited to my rather unpredictable life as a wandering foreign correspondent. A Leica M6, he said, and equipped with a remarkable lens,

---

* My late father entirely approved of the purchase of this camera, as Johann Voigtländer's company, though originally Viennese, had moved during the political turmoil of 1848 to the German city of Braunschweig, in Lower Saxony, and my father felt a long-standing affection for this city, despite (maybe because of) his having been incarcerated there as a prisoner of war during the closing months of World War II. "Damn good engineers, those Saxons," he harrumphed, and handed me ten pounds, as I remember, for the camera that set off a lifetime loyalty to 35 mm film and format. Voigtländer lenses made in the late nineteenth century were all constructed to the greatest mathematically determined precision, and were very fast and highly accurate—it remains one of the tragedies of German photography that this pioneering firm had to be wound up in 1972. Cameras and lenses are still made under the Voigtländer name, but under license by a Japanese company.

a (then-unfamiliar to me, but already legendary to those in the know) little black cylinder of robust delicacy, the phenomenally light and extraordinarily fast confection of air, glass, and aluminum known as the 35 mm f/1.4 Summilux.

That little lens stayed with me, performing journeyman work for the newspapers and magazines for which I worked, for more than a quarter of a century. It then went on to serve briefly on a newer and very different Leica body that I acquired much later. Eventually, I succumbed to the advice of my betters and bought that lens's natural successor, the 35 mm f/1.4 Summilux ASPH, which had an aspherical lens with what is called a floating element to it—regarded at the time of this writing as perhaps the best general-purpose wide-angle camera lens in the world, and probably the classic popular exemplar of high-precision optics.

There are certain ineradicable truths in the world of optical hyperprecision, and one of them, by near-universal agreement, is that the best Leica lenses are and long have been of unsurpassed quality, and deservedly represent the cynosure of the optical arts. The century-long arc of progress began with the moment in 1913 when Oskar Barnack—legend has it that he was asthmatic, and needed a lightweight camera—made both the first 35 mm film and then the first-ever Leica camera, called the Ur-Leica. It led to the creation of the supremely good lenses of today, a trajectory of progress in optics that mirrors much of the progress of precision more generally, even though using materials that, unlike most of the devices in this account, are invariably, and for the best results, transparent.

The optical journey itself begins almost a century earlier still.

If humankind's acceptance of light and dark began the moment the first eye was opened, or blinked, or shut, then the first questioning of optical phenomena probably started soon thereafter. The nature of shadows, of reflections, of rainbows, of the bending of sticks in pools of water, of shades and tint and hues of color—all would have come first, and then later there would have been considerations of the action of mirrors, of burning glasses, of the twinkling of stars and the steady light of planets, of the anatomy of the eye—all inquiries that are recorded in writings (Greek, Sumerian, Egyptian, Chinese) from at least three thousand years ago. Euclid's *Optics* was written in 300 BC, and though it is mainly a treatise on the geometry of angular perspective, and the belief that light to the eye is created by an ether-like substance called "visual fire," it laid the groundwork for Ptolemy's theories of five centuries later, brought some

The prototype Ur-Leica, fashioned in 1913 by the Leitz employee Oskar Barnack. It was small, light, with a near-silent shutter, and a 24 × 36 mm film format.

detachment and sophistication to the science of astronomy, and advanced theories of refraction and reflection that have not changed much to this day.

Surgery on eyeballs had already revealed the existence of a lens, a *perspicillum*, which, from its secure position at the front of the iris, magnifies all that it sees. It was a Swiss doctor who first displayed the lens of the human eye, and gave it the name that Romans had for centuries given to the small pieces of glass that the optically afflicted used for helping with their poor vision: *perspicillum* in later years denoted either a telescope, to see distant things up close, or crudely made

Even though Ernst Leitz famously helped his Jewish employees to leave Germany in droves, his cameras were much used by Hitler's military. Here are a pair of IIIcs worn by a Kriegsmarine seaman.

and ad hoc spectacles, which helped make close things appear sharply in focus and the illegible capable of being read.

Nero, myopic in more ways than one, was said to have watched gladiatorial contests through a conveniently curved emerald. The first true spectacles appear in images drawn in Italy in the thirteenth century, with simple lenses maybe, but life-changing for those who required them or for uses that allowed for the discovery of the distant unknown. Then came Galileo, and Kepler and Newton, and theories of light became ever more complex, and the exactitudes of geometrical optics took over from a hazy belief in visual fire; and then telescopes and binoculars and microscopes were made; and Benjamin Franklin reputedly created bifocal lenses, the glass more convex in his spectacles' lower half for reading and less rounded above a metal spacer, and so allowing for viewing at distance, in the early 1780s or, maybe, according to new research, as much as fifty years before that. Finally, in due time, with the realization of light sensitivity among various families of chemicals, the scientist and inventor Nicéphore Niépce snapped the first photograph and preserved one modest illuminated moment (even though it was a moment no less banal than its title, *A View from the Window at Le Gras*) for all time.

Snapped is hardly the word. Niépce used a camera obscura, at the back of which he mounted a pewter plate he had painted with a thin layer of a kind of bitumen he had discovered would harden upon exposure to light, becoming less hard in those places where the lens directed the lesser light and firmer where the illumination was intense. The asphalt was also selectively soluble—it could be washed away with a mixture of lavender oil and white gasoline—and Niépce realized with decisive logic that the firm parts would likely be more resistant to washing and the softer parts easily swept away. So, using this kind of chemical reaction to light and dark, Niépce took a photograph. It was a crude picture of a rooftop terrace made of blocks of stone, with a grove of trees at center stage and, across slightly to the right, a distant horizon with steeples and vague outlines of hills. It is barely recognizable, yet it is undeniably a vague image of just what his primitive little camera saw.

The picture was taken in the summer of 1826, in an east-central French village named Saint-Loup-de-Varennes (now a place of pilgrimage among the world's photographers), and with an exposure time of many hours, perhaps even many days. There is nothing either precise or accurate about the image, though it has a strangely ethereal beauty to it, and is viewed with great and deserved reverence in a vitrine in a much-protected vault at the University of Texas at Austin.

We know less than we might wish of the kind of lens Niépce employed on that long-vanished sultry summer's day—was it made of rough or polished glass, of ground crystal, or of a piece of amber found in a riverbed? We can suppose, but we cannot be sure. It was certainly fixed solidly in the camera box, and was certainly composed of just a single element, a single transparent entity. It was probably lemon shaped, convex on both sides. From examining the image that resulted, we know that it suffered from all the classic limitations of early photography: an inability to focus being one, an inability to capture sufficient light another, with distortions at the edges and at the sites where more light was falling. It certainly had no pretensions to being precise. Yet it is quite rightly a piece of deliberate creation, the haunting nature of its imagery a foreshadowing of a whole new art form to come.

Since Niépce's pioneering work, lens designers have discovered a host of technical problems that can conspire to spoil a photographic image: chromatic aberration, spherical aberration, vignetting, coma, astigmatism, field curvature, and problems with bokeh* and the so-called circle of confusion being among the best known. They have therefore experimented endlessly to produce compound lenses of great complexity

---

* From the Japanese word for "blur," *boke*, or the "quality of blur," *boke-aji*, bokeh is nowadays a much-courted aspect of optical quality that deals with how a lens manages the out-of-focus parts of an image—whether it renders them in an attractive or an ill-considered manner. The fact that modern photographers are so fascinated by bokeh is a reminder that lens sharpness is most certainly not the most valuable quality of a good lens—lightness, versatility, speed, and bokeh are all of greater moment to photographic artistry than the ability to make a picture filled with fine detail. The *circle of confusion* is a term of art and a related topic, dealing with the precisional aspect of photographic depth of field.

that correct for all these trials but that are at the same time fast and light and pure and true, and that contrive to make images that are as close to technical perfection as it is possible to imagine. The 134-year journey from Niépce's creation in 1826 to the designers and makers who created Leica's first 35 mm f/1.4 Summilux lens in 1960 offers a demonstration of a great optical trajectory, from simplicity to high precision, marking a passage in time from which all images were necessarily vague, to today, when, if desired, all can be razor sharp—not necessarily more beautiful, but forensically useful, in which highly detailed and accurate records of moments in time can be produced and preserved, and which, because of their accuracy, are entirely amenable to being blown up many, many times.

The way in which this was achieved has as much to do with mathematics as it has to do with materials. Mathematical concepts such as angles are crucial—angles of refraction, for example, or angles of dispersion, both of which are determined in large part by the kind of glass used in a lens. Refraction is a measure of how much a lens bends light, dispersion of how varied are the angles at which a lens refracts light of different wavelengths (that is, of different colors). Early lens designers did their best to limit spherical aberration and chromatic aberration (the very visible consequences of too much refraction and too much dispersion) by the brilliant idea of grinding two lenses of different materials such that they fitted exactly together—and in doing so, in the late 1830s, they created the first kind of multi-element lenses.*

The multi-element arrangements that followed, and that have dominated fine lens making ever since, began primitively enough, with just the two lenses pressed together. In these early examples, one lens would be made of a glass with specific refractive properties, such as so-called crown glass, which has a very low refractive index; and the other would be made of so-called flint glass, which has a very different chemistry, a high refractive index, and very low dispersion. Grind them into complementary shapes and press and cement the two together, and you come up with what is called a doublet.

The illuminated image whose reflected rays pass through this doublet are then focused onto the film at

---

* Niépce and his cohort stuck with just one element, probably glass, and initially simply convex on both sides. Two years after his first experiments with asphalt and lavender oil, he did, however, warm to the use of meniscus lenses, which had a concave side facing out and a convex side closer to the film. Niépce also worked to keep the pinhole of his camera obscura very small and centered on the lens so that only its nonaberrant center would be employed to collect imagery and so make pictures.

the back of the camera in a manner that will be much more disciplined, focused, and lifelike than the fuzzy, blurred-edge, and randomly aberrant imagery previously offered by single-lens cameras of yore. The crown glass lens deals with one problem, the flint glass lens with another—and the two together are ground so perfectly that, optically, they act as one, with one physical effect on the light, variously now tinkered with by its two components.

Multi-element confections of one kind or another have dominated good-quality camera lens designs ever since. Optics designers are today rather like orchestral conductors, maestros who marshal and corral morsels of carefully shaped and exquisitely ground glass of varying chemistries and optical properties into configurations that will provide the most harmonic and pleasing management of light for the task the lens is designed to perform. Lens geometries are infinitely variable, lens materials even more so—tiny additions of rare earths change the dispersion and the absorption and the refractive abilities of transparent materials, while certain nonglass materials (germanium, zinc selenide, fused silica) perform particularly well with certain kinds and wavelengths and intensities of light.

The job of a lens is to capture the light and present it to the camera and the film or the sensor it holds. As cameras and films and sensors became ever more able (allowing for higher shutter speeds and finer grains and, in the digital world, ever more pixels), the manufacture of light-presenting lenses became ever more demanding, the arrangement of glasses within ever more intricate. Portrait lenses, for example, had one kind of configuration: an early kind had four lens elements, two cemented together, two grouped together but with air sandwiched between them. Lenses designed for capturing photographs of landscapes, for their part, had very different arrangements, as did wide-angle, close-up, telephoto, macro, fish-eye, and zoom lenses. Indeed, some variable zoom lenses have as many as sixteen elements, some of them movable, some of them fixed, some stuck firmly together, and some separated by distances large enough, but nonetheless very accurately measured, for the resulting lenses to be of bewildering and barely manageable lengths, often needing a tripod support of their own, with the camera body a mere bagatelle fixed to one end.

Leica—the name is a blend of the company founder's surname, *Leitz*, and his product, a camera—entered the field of exact optics in 1924. The inventor of the first 35 mm camera, Oskar Barnack, whose two Ur-Leicas were built in 1913 and whose O-series production camera was offered to the public in 1925 (the interlude being due, of course, to the Great War), was incredulous at the quality of the early lenses. The O series was equipped with a lens designed by a long-forgotten optical genius named Max Berek. It had five glass elements (a cemented triplet and two singlets), and when Barnack saw the results, a clutch of eight-by-ten-inch prints sent to him in the mail, he dismissed them out of hand: they couldn't possibly be the enlargements of the 35 mm images he had been promised. Yet, of course, they were—blown up tenfold and losing none of their crispness in the process. The lens that took the images went on the market as the 50 mm Elmar Anastigmat, and it remained a classic for generations, and is a priceless collector's item today.

And down the years, so the lenses processed, all with code names ineradicably linked with Leica: the Elmax, the Angulon, the Noctilux, the Summarex, the too-numerous-to-count Summicrons, and the bijoux of the family: the three focal-length superfast lenses (35 mm, 50 mm, and 75 mm) that were given the code name Summilux, and all of which were designed to offer the most stringent accuracy even at their widest aperture of f/1.4.

For the making of all these, the common Leica standards were unparalleled. Whereas most camera makers work today to an industry standard of 1/1,000 of an inch, and with Canon and Nikon working their mechanicals to a supertight 1/1,500 of an inch, Leica bodies are made to 1/100 of a millimeter, or 1/2,500 of an inch. And with lenses, the tolerances are even tighter. The refractive index of Leica optical glassware is computed to ±0.0002 percent; the dispersion figures (the so-called Abbe numbers) are measured to ±0.2 percent, against an industry-agreed international standard of 0.8 percent. And the mechanical polishing and grinding of the lenses themselves are performed to one-quarter lambda, or one-quarter of the wavelength of light, with lens surfaces machined to tolerances of 500 nanometers, or 0.0005 mm. And with the aspherical lenses that cut so markedly down on the tendency at wide apertures to display spherical aberrations, machining of the glass surfaces is performed down to a measurable 0.03 micrometer, or 0.00003 mm.

The lens I now have as successor, the mighty miniature classic 35 mm f/1.4 Summilux, ticks all these boxes, insofar as it now comes with one aspherical lens element that has, in a very recent iteration known as the aspherical FLE, four of its nine elements closest to the camera body *floating*, free to travel together as one within the lens structure, giving the most memorably good results. This lens has become perhaps the best-regarded wide-angle piece of optical glassware ever made, by anyone: the reviews have been stellar.

To hold one of these lenses (a scant ten ounces of aluminum, glass, and air) is to hold almost the most precise of modern consumer durables—with one notable and rather obvious exception: the smartphone. Within that particular handheld device (as I shall outline later) is a sturdy mix of mechanical exactitude, the various component parts finished to the severest of tolerances. Yet there is also, and essentially, a mass of electronic precision to it, a gathering of myriad components where no moving parts are present to interfere with what is designed to be their constant perfect performance. The making of the circuitry that runs the smartphone, and similar versions of which run other devices big and small that profoundly affect so very much of today's lives, takes the concepts of accuracy and precision into a whole new realm. But that is for later.

Mechanical precision, at this high and demanding level, can on occasion stumble, however—tiny errors can be made; they can accumulate, resonate, and harmonize to become major errors, after which they can become the origin of problems that the designers may never have supposed or imagined.

For instance, those workers in Hucknall, Nottinghamshire, who in 2009 incorrectly machined the tiny metal tubes made to lubricate the turbine section of a jet engine never would have imagined that, a year later, because of their minuscule mistake, a fire would break out, the engine would destroy itself, and the lives of almost 470 people would be briefly in the balance a mile up in the sky above Indonesia.

The kind of tolerances demanded in modern precision devices allow essentially for no errors, but insofar as human beings are still involved in the manufacturing of precise things, human failings do occasionally creep into the process. Possibly the most recent classic example of an imprecise human failing intersecting with a mechanism made for a precise and uninhabited world was that exposed for all to see with the launch, and then the failure and, finally, the great success, of the Hubble Space Telescope.

"Ask any person the name of a playwright," remarked Mario Livio, a NASA astrophysicist and senior scientist on the telescope project, "and most of them would say Shakespeare. Ask them the name of a scientist, most of them would say Einstein. Ask the name of a telescope—they will all say Hubble."* There is a dis-

---

* . . . though some in Europe might well rather cite Herschel. Few today can possibly gainsay the astronomical achievements of this remarkable telescope-wielding family, three generations descended from an oboe-playing soldier and former gardener from

tinct public reverence for this telescope, in part because of the sheer magnificence of the images it has sent back to Earth from space in recent years. To some of us, though, Hubble is perhaps also regarded fondly because of its vulnerability, because of its troubled story, its phoenixlike rise from the ashes of its awful beginnings.

It was placed gently into orbit, 380 miles above Earth, on April 24, 1990. At the time of its launch, Edwin Hubble, America's great deep-space astronomer, who was the first to suggest that the universe might be ex-

---

Germany, but who themselves fled to and settled and performed most of their stargazing in England. William and Caroline Herschel, siblings, were the first to win fame—he for discovering the planet Uranus in 1781; she—hitherto a quite uneducated maidservant—for helping her brother discover more than a score of comets and some twenty-five hundred nebulae. The image of the pair spending nights grinding and polishing lenses and mirrors to such a degree of precision as was attainable in the mid-eighteenth century lingers still in the annals of astronomy's charms. William's son, John Herschel, was to become so revered a scientist—polymathic but supreme in sky-searching—that he was buried in Westminster Abbey beside Sir Isaac Newton (the common man may bless him for his interest in cameras, and his invention of the terms *positive, negative, snap-shot,* and *photographer*). He was also happily fertile: his fifth child (of twelve) and second son, Alexander, was himself no mean astronomer, and would become a professor, a Fellow of the Royal Society, and a leading authority on meteorites.

The Hubble Space Telescope was launched and then placed
into orbit 380 miles above Earth on April 24, 1990. After
its mirror was found to be flawed, it was repaired in space
in December 1993. The Hubble has since performed near-
flawlessly, sending back countless captivating images of
interstellar space.

panding, who examined the universe beyond our own
pitifully small galaxy, and after whom the telescope is
named, had been in his grave for almost forty years.
This in-space observatory, at launch time more than a
quarter century in the planning, was in essence a project to push still further inquiry into the state of faraway
stars and galaxies and nebulae and black holes.* The

---

* NASA has now all but completed a vastly more powerful (and
at eight billion dollars, much costlier) device, the James Webb

device was less a memorial to him than a continuation of his work.

The orbiter *Discovery* took the Hubble up into space, to a place well above the distortion and pollution of Earth's atmosphere and comfortably away from the harsher pulls and tugs of geomagnetism and of gravity. It was the seventh flight of this particular workhorse of a space shuttle—the orders were for a short (five-day) in-and-out, drop-it-and-leave-it mission. The flight was designated STS-31, and, despite the numbering, was the thirty-*fifth* mission of NASA's five reusable Space Transportation System vehicles—except that, at the time of the launch, the number of orbiters was down to four.

And because of that grim arithmetic, those who watched the shuttle launch on that warm Florida spring morning were more white-knuckled than usual. Four years before, the sister orbiter, *Challenger*, had exploded

---

Space Telescope, which is due to be launched from the European spaceport in French Guiana, in April 2019. The telescope will float almost a million miles from Earth, well beyond the reach of any shuttle-hoisted repair teams; consequently, its manufacture, and the planning of the deep-space maneuvers that have to be successfully accomplished before it can make a single observation, are being rehearsed over and over, to ensure that everything works down to the finest detail.

seventy-four seconds after takeoff, killing everyone aboard. After three years of memorials and investigations and repercussions and modifications, NASA decided that *Discovery* would be the vehicle to perform the first post-accident flight. Her mission, then, in September 1988, was designed as much for confidence building as it was for the execution of major science. The nation breathed a collective sigh of relief when the launch in Florida, the four subsequent days spent soaring around the world, and the then-picture-perfect landing in California all went without incident.

Discovery flew again, twice, in March and November 1989, by which time the country had become largely convinced that the problems that had brought down *Challenger* (a rubber seal that had stiffened in the sub-zero weather of the midwinter launch day and caused fuel to leak from the solid rocket booster) had been solved. Still, this STS-31 was an extremely high-value mission: the Lockheed-built telescope and its Perkin-Elmer Corporation optics, snugged safely away in the cargo bay, had already cost the taxpayers around $1.8 billion. There was much consequent anxiety before the launch. This anxiety barely diminished even after the successful liftoff. Indeed, the pressure was unrelenting until the moment the next day when the crew used *Discovery*'s Canadian-made robotic arm to lift the bus-size payload out of the hold; set up its solar panels and telemetry and radio aerials; *switch it on*, as it were; and finally let the first of NASA's so-called great observatories free-float into orbit.*

Hubble, enormous during its construction (the size of a five-story house) but minuscule in appearance when floating in the immense nothingness of space, is perhaps not the prettiest of sights. There is something rather awkward, almost teenager gawky, about its appearance—like a once-chubby, silver-coated boy who has suffered a sudden growth spurt and who floats along

---

* The four observatories, were they to act in concert, seek to observe the universe through a considerable portion of the electromagnetic spectrum—Hubble, the best known, investigating all the way from ultraviolet to the near-infrared, by way of the entirety of the visible spectrum. The Compton Gamma Ray Observatory, sent into orbit aboard a shuttle in 1991, looked at violent and high-energy events out in space, those that emitted bursts of gamma rays. In 1999, the Chandra X-ray Observatory, also dropped from a shuttle, looks at X-ray emissions from black holes and quasars. Finally, in 2003, a Delta rocket took the Spitzer Space Telescope up into high solar orbit, from where it observes thermal infrared emissions, invisible from Earth because the very short wavelengths (as low as three microns) cannot penetrate Earth's atmosphere. The Compton Gamma Ray Observatory reentered the atmosphere and is no more: the remaining three are still working like a charm.

on his own looking, his mother unable to afford new clothes for him, rather wrinkled and ungainly and uncomfortable with his new shape. Moreover, the hinged lid at one end, which admits the light into its main tube, looks awkward, too—so much like the open top of a kitchen garbage can that you rather expect a foot pedal to be protruding from somewhere, keeping it open all the while. Instead of pedals, there are solar panels, squared off and able to furl and unfurl as the temperature varies depending on the telescope's attitude and position.

It is wanting in prettiness maybe, but its two builders and NASA, their customer, knew it to be an extremely powerful piece of kit. In many ways it was a quite simple telescope, a so-called Cassegrain reflector, well known to any amateur backyard stargazer, with a pair of mirrors facing each other—the primary mirror gathers the light and reflects it to the smaller, secondary mirror, which then reflects it back once again, through a hole in the center of the primary and to a variety of observing devices (cameras, spectrometers, and detectors of various wavelengths from ultraviolet all through the visible-light spectrum and into that of the near-infrared). The detectors were packed into telephone booth–size boxes arranged tightly behind the primary mirror, and from which the gathered data would then be beamed as telemetry signals back down to Earth.

The specific design of Cassegrain used in the Hubble telescope involved the use of a particular mirror shape—they were called hyperbolic reflectors—that specifically reduced the chance of two types of aberrations in the image, the comet trail–like aberration known as coma and the edge-of-lens error known as spherical aberration. And as Hubble settled itself into space that May (once *Discovery* had fired her braking thrusters and dropped out of orbit and spiraled her way down earthward to leave the telescope quite silent and alone), the device, with all its optical distortions and aberrations duly thought of, anticipated, and averted, seemed richly pregnant with astronomical potential.

Except that, six weeks later, an unanticipated nightmare began to unfold, an unimagined nightmare—for this was no *Challenger*, where frantic engineers far away, who knew of the risks of launching in freezing weather, tried desperately to cancel the flight. In Hubble's case, all was blissfully normal, everyone lulled into a state of contentment—and, as it happens, hubris.

All began routinely. Three weeks after the telescope had reached orbit, on May 20, by which time all were confident that it had cooled itself from the warmth of the Florida beachfront to the ambient temperature of its new surroundings, Mission Control sent out a signal to unlatch the hinged front door to the optics.

Hubble was now open for business. The first light from a million stars—and this was the way the moment was named, the First Light—flooded into the barrel of the telescope. From there, it headed toward the primary mirror and made its reflective journey back and forth until, ultimately, it fed itself into the detectors and became the data that were so eagerly awaited by the watchers of the skies on Earth at the Space Telescope Science Institute at Johns Hopkins University in Baltimore, far down below. The transmissions were perfect. Data were streaming down, just as they should. An astronomer named Eric Chaisson inspected the incoming images, and then, suddenly, as he examined what he saw, he felt, as he put it, "a total deflation in my gut."

Something was horribly, horribly wrong. Everything was blurry.

Two weeks later, Edward Weiler, who at the time was chief scientist for the Hubble program at NASA's Goddard Space Flight Center, thirty miles away, was basking in the apparent success of the early stages of the mission. But then he received an alarming telephone call. It came from one of his colleagues at the science control room in Baltimore. Try as they might to improve matters, the panicky-sounding scientist there told Weiler, every single picture that had been beamed down from Hubble was totally out of focus (except, by some cruel trick, the very first, which seemed very sharp indeed).

They had tried for days, not daring to report the news, to fine-tune the images by moving the secondary mirror by infinitesimal amounts, to tease out a picture from the primary mirror, which was sharp and clear. But while most of the astronomers in the control room agreed that the image qualities they acquired were as good as or better than their equivalents from ground-based telescopes, they were not nearly as good as they should have been. Indeed, not even that was true; it was wishful thinking. The brutal truth was that not a single one of the pictures could be coaxed into a usable degree of sharpness. Each one of them was a grave disappointment. They were worthless, useless. The mission, by all accounts, seemed all of a sudden to be judged an abject failure.

The ghastly news was broadcast to the world on June 27, 1990, two months after the launch. The vision of a gathering of NASA bureaucrats in their business suits, all sporting long faces of gloom (Ed Weiler, blond and in those days quite cherubic, as downcast-looking as the rest of them), lined up to face a corps of incredulous reporters, each of the scribblers holding an image of interstellar wreckage before him or her, will linger long in the memories of all who watched the awful moment of admission. It was all true, the gathered men said, some of them barely able to get the words out. The eight-foot-diameter primary mirror of the telescope, though at the time the most precisely made optical mirror ever built, appeared to have had its edges ground too flat.

It was out by only the tiniest amount, a fiftieth of the thickness of a human hair, but that was enough to wreak optical devastation. The coma and spherical aberration caused by this one tiny mistake rendered almost all the observations valuelessly befuzzed, with distant galaxies looking thick and edgeless, like marshmallows; stars looking like powder puffs; nebulae, like the merest patches of random discoloration. Images as mediocre as this might as well have been gathered by someone with an eight-inch telescope in a smoky backyard in Ohio. Indeed, there seemed no need to have spent almost two billion dollars and twenty years' worth of the labor of men and women in America and Europe (for this was a European Space Agency venture as well as NASA's) and beyond.

The press was viciously unkind. Many agreed that the Hubble was a device no better than the infamously unpopular Ford model the Edsel. Maybe the telescope had been designed by the myopic cartoon character

Mr. Magoo. Lemons in space were spotted by many newspaper cartoonists, as was static, the television post-broadcast snow being all that NASA seemed to have discovered, its meaninglessness filling the Hubble-known universe. NASA appeared to be in the business of building "technological turkeys," said one angry Maryland senator. The optical catastrophe may have killed no one, but in terms of national embarrassment and humiliation, the error was reckoned by some of the more excitable politicians to be on a par with the crash and fire of the *Hindenburg* and the sinking of the *Lusitania*.

Indeed, in the view of some more rabid legislators, who, after all, held the NASA purse strings, the ruined performance of what was the costliest civilian satellite ever built—and there were now other mistakes showing themselves: faults with the solar arrays made the whole telescope shake and shimmy, lowering expectations for much serious and successful science—put the future of the entire agency at risk. Only four years before, *Challenger* had exploded because of agency ineptitude. Now this. Twenty-five thousand NASA employees, uncounted thousands of contractors and suppliers—all suddenly seemed to have their futures on the line.

It turned out to be all down to one company, at the time called the Perkin-Elmer Corporation, which was based in Danbury, Connecticut, ninety minutes' drive north of New York City. The firm had ever since the late 1960s ground the mirrors and made the cameras for a series of highly classified spy satellites. It was a well-experienced major player on "the dark side," that mysterious shadowland of research and manufacturing for the American military, whose role in all things precise is acknowledged but seldom spoken about in detail. A windowless cement building on a hill outside Danbury housed the polishing and grinding machinery that had for years enabled the army and the navy and the various spy agencies to look down from on high into forests and fields and bases and houses all over the world, and gain knowledge without anyone below ever knowing they were doing so.

Come 1975, and Perkin-Elmer won a new contract: seventy million dollars, a deliberate lowball bid,* to

---

* Monetary matters are rather beyond the scope of this story, except that, to this day, employees of the shamed company insist that corners cut because of a lack of money were a principal cause of the mistakes made. NASA was uncertain that the firm could do the job for 70 million, but agreed to the lowball offer—the firm underbid Kodak, for instance, by a stunning 35 million—and winked at the arrangement, saying they could wheedle any additional funds from Congress later. Yet, later, Congress balked, and Perkin-Elmer had to try to make the mirror with the money it had, and of which it had demanded so little for the sole purpose of winning the contract and so enhancing

shape and grind and polish the primary mirror of a giant new telescope. The enormous blank glass disk was delivered from the Corning glass factory in the fall of 1978. Right from the start, the auguries were none too good. A quality-control inspector almost fell onto the glass, saved only when an alert colleague grabbed his shirttail. The joining of the three component parts of the optical "sandwich" that would make up the mirror blank went badly wrong: the 3,600-degree furnace fused the internal structure in a way that would probably cause it to crack during polishing, and so, for three months, Corning workers had to slice out the fused portions with acid and dental tools.

Never before had Corning made so challenging a piece of glass. Never before had Perkin-Elmer been given so demanding a remit: the NASA contract required the firm to grind and polish the finished fused-quartz glass piece, taking away at least two hundred pounds of material in doing so, and shape the immense tablet into precise convexity, with a surface of a smoothness never achieved or desired before. No part was to deviate by

---

its reputation. As we now know, exactly the opposite happened: the firm's reputation was left in tatters, and it had to pay NASA a hefty sum in compensation for its incompetence. It has changed ownership twice, and is now part of United Technologies.

more than one-millionth of an inch. The satin-smooth surface was to be such that if the mirror were the size of the Atlantic Ocean, no point on it would be higher than three or four inches above sea level. If it were the size of the United States, no hills or valleys on its surface would deviate by more than two and a half inches from plane.

The crude grinding of the glass slab began at the Perkin-Elmer plant in Wilton, Connecticut, just as soon as Corning delivered it, yet all manner of delays plagued even this period in the mirror's history—especially the so-called teacup affair, when a teacup-size web of internal cracks and fissures was found deep inside the glass, and had to be cut out and reamed and remelted, in a process akin to brain surgery. Finally, in May 1980, already nine months late, but with the mirror's basic shape achieved, the great glass object was carefully trucked to the hitherto secret facility outside Danbury, and the serious polishing began.

The piece was carefully lowered onto a fakir's bed of 134 titanium nails, a crude simulation of the gravity-free environment in which Hubble would eventually operate, and a computer-directed swiveling arm was moved into place over the piece. A spinning cloth pad at the arm's end, smeared with a variety of progressively less and less abrasive substances (from diamond slurry to jeweler's rouge to cerium oxide), was then lowered onto the face of the glass plate. Under computer control, it steadily stripped away and purged and polished and smoothed the surface, with polishing runs lasting for as long as three days, day and night. Polishers often worked back-to-back ten-hour shifts. Three days of polishing would be followed by a move to the testing room. Then, based on what the testers measured, new

The Hubble telescope's eight-foot-diameter primary mirror being polished at the Perkin-Elmer Corporation's top-secret facility in Danbury, Connecticut. An overlooked measurement error on the mirror amounting to one-fiftieth the thickness of a human hair managed to render most of the images beamed down from Hubble fuzzy and almost wholly useless.

computer instructions would be issued, to polish this segment at this pressure and with this abrasive powder for so many hours and to then polish that section at quite another pressure and with a wholly different abrasive for more or less a similar period of time. That run would be finished three days on, and new tests would be ordered, and so the routine would continue, week after week after week. Testing was usually done at night, so as to minimize vibration from the daytime cavalcades of trucks passing by on Route 7; managers switched off air conditioners, too, for the same reason. Companywide, all were most scrupulous, firmly believing in their reputation for paying attention to the most minute of details.

Yet, once in a while, they made trivial mistakes, or rather, they wrongly instructed machines, which then made their own trivial mistakes, as ordered and on demand. Gone were the days when a skilled mirror maker could make certain of the precision of a surface by running a practiced thumb along it. Now such measuring was all done by machines, and one day, a Danbury engineer punched the number 1.0 into a terminal instead of 0.1—and watched in horror as the abrasive tool started to gouge a trench in the side of the glass. Mercifully there was a check technician standing by holding a Kill switch. He noticed the incipient gouge and stopped the polishing dead in its tracks. The small nick in the glass never fully went away, but it was smoothed over to a degree and left as a reminder, one that properly informed astronomers could work around.

It was in the testing room that the fatal error was made, and it was not a trivial one. For, while the smoothness or surface precision of the mirror face was being created with unforgiving certainty, the measurement of it was entirely and utterly wrong. The Danbury crew had made their measuring tool incorrectly: they were using an instrument that was rather like a straightedge that was stated to be, and was thought by everyone who used it to be, exactly one foot long, but that in fact measured thirteen inches—and nobody ever noticed. This was unknown to the engineers—they were busily measuring and then manufacturing something that was perfect but entirely wrong. They were making a telescope mirror that would be *precisely imprecise*.

The tool they made to measure the glass was a familiar piece of equipment called a null corrector. It was a metal cylinder about the size of a beer keg, and it held a pair of mirrors and a lens. Laser light was bounced against the two mirrors, then through the lens, where it would be directed to and bounced off the polished surface of the mirror before being passed back to the corrector's lens and mirrors once again, and to the point

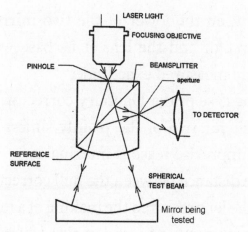

A worn patch of paint and three tiny washers turned out to be the culprits that led to a so-called null corrector giving a false result to the shape of the Hubble's main mirror.

where the light originated. If the polishing was perfect, then the light going out and the light coming back would match, wavelength for wavelength, and would produce in a photograph a pattern of straight and parallel lines. If the mirror was not the desired shape and smoothness, then the waves would interfere with one another, and the photograph would display an interference pattern. The null corrector, a million-dollar specially built measuring device, was in essence an interferometer, a device that, if properly set up, would be capable of confirming the absolute precision of the mirror surface, and to a fraction of the wavelength of light.

It could, that is, if—and this was a crucial if—the

distance between the lower of the two mirrors inside the null corrector and the lens at its base was known with precisely measured exactitude.

And in the case of the Danbury corrector, it wasn't. And it wasn't for two of the plainly silliest and most prosaic and imprecise reasons imaginable.

To set the distance between the null corrector's lower mirror and the lens required the making of a metal rod of the exact length required—and so three such rods (two as spares), made of the less-heat-sensitive alloy Invar, were made, measured, and cut. One of these metering rods was then fitted inside the null corrector and a laser trained onto its tip. Using a specially made microscope, a technician then worked with a laser interferometer to set the distance that the lens was to be adjusted to, so it would end up in the right place. It was a tricky job, but not an impossible one—and to make it easier for the worker, a special guide cap had been fitted to the very top of the metering rod with a minute laser-beam-size hole cut in it, indicating the place where the laser should be aimed, to ensure the technician hit the rod's very tip.

Crucially, fatally, the cap had been covered with a non-laser-reflective coating, to make quite certain the laser would focus not on the cap but on the metal tip that was visible through the hole alone. It turned out, though, that a small portion of the coating on the cap had worn off, and the laser focused on, and was reflected by, that part of the cap, instead of traveling on through the hole to the rod's metal and similarly reflective tip. The cap's surface was exactly 1.3 mm higher than the tip of the rod, so the laser interferometer calculated the distance incorrectly by that exact amount.

The difference then made it mechanically impossible for the technicians to set the lens where the laser had said it should be set. The bracket holding the lens was out by 1.3 mm. Something was needed to bring this bracket down by 1.3 mm. There was no time to custom-make a new bracket.

So, resourceful as technicians often have to be, they made a decision. They would put three household washers into the null corrector to force the tiny lens 1.3 mm lower. They had to do this because the laser could not possibly be wrong. Lasers are so precise they never lie. They tell the absolute truth, with cold-eyed reliability. So three washers, flattened with hammers such that together they would form a tiny sandwich 1.3 mm high, were placed above the lens, which then and at last assumed the position that had been ordered.

After which, handling with extreme, crown jewel–like care the now completed but, as it happened, now profoundly flawed null corrector, the technicians slid it into position above the telescope mirror. Employing its electronic infallibility, the engineers measured and measured again, and eventually proved to their own satisfaction that in size and shape and configuration the Hubble Space Telescope's primary mirror was exactly, precisely, as ordered by NASA.

Yet it wasn't. It looked good according to the null corrector, but the null corrector was wrong. The NASA inquiry was able to demonstrate this because the Perkin-Elmer mirror makers had left it in the testing room, and had left the testing room exactly as it had been when they made their final measurements on the completed mirror, nearly a decade before.* The result was that at the edges of the mirror the tiny error in the metering rod, and thus in the null corrector, had produced a change in the measurement of the primary mirror's shape that amounted to a 2.2-micron flattening around its edges—the famous one-fiftieth-of-the-thickness-of-a-human-hair deviation from design. A literally microscopic error, but one that resulted in the wholesale uselessness of the images

---

* It is easy to forget that many years went by after the mirror was completed—the *Challenger* disaster and innumerable technical delays in the manufacturing of the rest of the Hubble telescope pushed the launch date back, and back, and back. Meanwhile, the mirror system was kept in storage at Lockheed.

sent down from space early in that summer of 1990, and which rendered the Hubble a laughingstock.

"If you had polled all the engineers and scientists at the Cape the night before launch for the top ten concerns they had," remarked Ed Weiler some while later, "what could break on Hubble or what wouldn't work on Hubble, I would bet my house and a lot more that not one of them would put on their list [that] the mirror is the wrong shape and so we have now got spherical aberration. Nobody worried about that, because we were assured by the optics guys that we had the most perfect mirror ever ground by humans on Earth."

As indeed they had, but they also had an inaccurate measuring device that told everyone the mirror was perfect, and by its own standards, so it was. But its standards were disastrously imperfect, inaccurate, and wrong.

"For want of a nail . . ." goes the ancient proverb.* In this case, it was the simple want of a patch of paint on

---

* The saying was first brought to public attention in the mid-seventeenth-century anthology *Jacula Prudentum,* assembled by George Herbert, the saintly (and wealthy) vicar of the quaintly named church of Fugglestone St. Peter, a few miles from the cathedral city of Salisbury. The full proverb reads, "For want of a nail, the shoe is lost; for want of a shoe the horse

an Invar rod, coupled with a degree of insouciant carelessness among a group of harried technicians and their budget-strapped managers, that led not to the loss of a kingdom, of course, but to a cascade of events and risky ventures and the expenditure of yet more taxpayer money in order to make repairs.

For Hubble was in due course repaired, and made good. It has become so successful, in fact, that it is repeatedly cited as the most valuable scientific instrument ever made, allowing for exploration of the outer regions of the universe to a degree that astronomers had never dreamed possible. It did have its error reversed, and its defects repaired, and so well—yet this came about every bit as improbably as the making of the error in the first place.

The repair had to be performed out in space—there was no chance of bringing Hubble back down into the shop. The installation of corrective optics should have solved the main problem—it would have been rather like giving a severely myopic person a set of contact

is lost; for want of a horse the rider is lost; for want of a rider the battle is lost; for want of a battle the kingdom is lost." The same anthology also offered the suggestion, of a fierce man, that "his bark is worse than his bite." Hitherto they were believed equivalent.

lenses, or a form of Lasik surgery—yet, for a variety of technicalities, such a repair was going to be very difficult. The telescope tube was narrow and filled with a mess of instruments, pipes, and wires, and to send an astronaut swimming down into it with an oxygen pack and a wrench and a screwdriver and holding a new set of optical correctives was going to be exceptionally difficult, for a host of reasons.

Then one man solved this central problem, and he did so in a sudden moment of lateral thinking that came to him while he stood, stark naked, in a shower in a hotel bathroom in Munich, in the mountains in the south of Germany.

His name was Jim Crocker. He was, at the time, a senior Hubble optical engineer, and he was quite as devastated as the rest of the crew. Like most who had gathered in Germany for a crisis meeting of the European Space Agency, where everyone was imploring everyone else for a solution to the floating problems of Hubble, Crocker was obsessed with the need to make repairs. All that was needed was a means of inserting corrective optics, lenses or mirrors of one kind or another, into the stricken device. It would not be possible to put them in front of the primary mirror, between it and the secondary reflector, as not even the slenderest

Jim Crocker, a NASA optical engineer, was taking a shower in his hotel in Germany when he realized that a device basically similar to a German shower mount could be used to probe inside the Hubble telescope tube and repair the optics or install corrective ones. NASA agreed and sent up the necessary device, which resulted in the total instant repair of the stricken telescope.

astronaut known to NASA could slither into and out of the main tube. No, the only place to site the corrective devices—and there would have to be four of them: one for each of the detectors carried on Hubble—would be behind the mirror, in the detector space itself. But how to get them to deploy? That seemed the impossible bit.

Then Jim Crocker took his shower, and mused, as one does under a stream of hot water, looking idly at the shiny chrome-plated components of a typical German shower, and did a double take. He then looked more carefully.

The showerhead was held on a vertical inch-thick rod, on which traveled a clamp that held the head, which itself could be raised and lowered and locked in place to accommodate hotel guests of differing heights and preferences. Moreover, the showerhead that traveled up and down along with the clamp could itself be angled up and down and from side to side, depending on whether the occupant of the shower wanted to bathe his head or shoulders or wherever. The hotel maid had left the showerhead at the base of the rod, and had left it folded up, so that it was parallel to the wall. In order to use it, Crocker had to slide it up to his head height and then fold the showerhead outward to direct the water to his hair.

Why not, mused our drenched and ever-more-cleansed engineer shower taker, mount the telescope's corrective optics onto a rod like this? Why not have them folded flat as they were slid into position, to be extended automatically into their preplanned and precisely determined base locations, then unfolded, just like the showerhead, into the correct angles and exactly calculated places?

There would need to be five of them—five "showerheads" instead of one, each to service one of the five main instrument packages the Hubble carried. Making five would, in truth, be no more difficult than making one. Each would have the same function: it would intercept the beams of starlight that had been reflected from the secondary mirror and that passed back through the center hole of the ruined primary mirror. They would then act on these beams and, much like contact lenses or corrective glasses, would refigure, recompute, and refocus them so that when they passed into the various Hubble detectors, they would be as perfect and as sharp as if the misshapen mirror's misshapenness had never existed.

It seemed so simple a plan, and the engineers working on the fix jumped on it in an instant. All promptly set to work on creating a showerhead apparatus of their own, but one that would carry an array of tiny (dime- or quarter-size) mirrors into space instead of the customary shower of warm water.

As, indeed, they did. The device was to be named COSTAR, or the Corrective Optics Space Telescope Axial Replacement—"axial" because the part was to sit behind the primary mirror and work with the light that traveled along the axis of the telescope. Very basically, it was a telephone box–size container made to exactly the same specifications as one of the instruments already aboard Hubble, the least important of the four axial detectors, known as the High Speed Photometer, and which would now be sacrificed (to understandable howls of protest from its manager) to accommodate the box with the new foldout mirrors.

Engineers swarmed in to hand-make the COSTAR, and to make it exactly right—the ten mirrors (which, in the end, did not swivel out, as in Crocker's showerhead, but sat atop an extendable tower and then radiated out from it, horizontally) having to achieve positions correct to at least one-millionth of a meter in order to be able to intercept the rays from Hubble's two existing (and shamed) main mirrors.

One critical problem was how to make certain the beams of reflected light destined for the axial instruments missed quite another cluster of beams that were destined for the one instrument that was set not into the end of the Hubble but into one side of it, and which was itself being totally replaced for its own mirror-related problems. This was the immensely costly Wide Field and Planetary Camera, which was built at the Jet Propulsion Laboratory in Pasadena. It looked like a large slice of cake (though the size of a grand piano) and was slotted into the curved side of the Hubble. The astronomers had always reckoned on replacing this device with a new and improved one during one of the five planned shuttle servicing missions. Now, with the first such mission due, they could do two crucial things at once: replace the High Speed Photometer with the COSTAR package and also install the replacement Wiffpic, as JPL's Wide Field Camera was affectionately known, and which had corrective optics of its own built in, to compensate for the errors in the primary mirror.

All that remained to bring the saga to a satisfactory conclusion was for astronauts to go up into space to make the necessary repairs. Hubble should then have been totally fixed and should have ended up every bit as valuable a piece of astronomical equipment as had first been promised—which would happen if the servicing mission worked as it was designed to, and in particular, as long as no one during the servicing dared to touch, even lightly, the little mirrors on either the COSTAR or the replacement Wiffpic, as touching them would take the repaired Hubble right out of focus yet again.

It was *Endeavour** that was chosen for the critical voyage, for what was designated by the shuttle teams as STS-61 and by the Hubble teams as HSM-1, the first Hubble Service Mission. She was launched in the heat

---

* *Endeavour* is spelled in the British style because it memorializes Captain James Cook's flagship. The name of the orbiter (built to replace the lost *Challenger*) was chosen through a national school contest, which was won by classes based in Mississippi and Georgia.

of the Florida night just before dawn on December 2, 1993, with plans and equipment (including some two hundred specially made tools) destined to bring to an end the forty-four-month nightmare of this half-blind and shuddering telescope that was relentlessly orbiting near-uselessly around the globe. The Wiffpic and the COSTAR were in the cargo bay; spacewalks of exhausting duration were planned to make the necessary fixes; the incoming astronauts who were certified to venture outside knew there were thirty-one foot restraints and two hundred feet of handrails already built onto the Hubble, and they had brought their own as well, together with numberless tethering lines, to make sure that no one and no equipment would be lost to float away into the eternal nothingness.

The crew, using powerful binoculars, spotted the telescope on the third day of their mission. They then closed in on it with infinite slow care, extended the Canadian robotic arm when they were sixty feet away, grabbed the thirteen-ton (but out there, featherlight) device, and hauled it gingerly into the shuttle's cavernous cargo bay. The crew of seven then began a series of walks in space (extravehicular activities, as NASA still unimaginatively calls such things) to perform the various tasks they had been assigned. Walk One (EVA One, more accurately) involved replacing the three (of six) gyroscopes that had gone awry, but it also served to allow the team to get accustomed to the size and scale of the patient on which they were working. (They had been training for eleven months, performing all these various tasks underwater, to simulate somewhat the lack of gravitational pull in space.)

Walk Two had two of the astronauts repair and replace the telescope's damaged solar arrays, which were said to have been causing the shuddering of the Hubble—a rapid movement that hardly helped the out-of-focus situation, yet paled in insignificance by comparison with the central problem. Matters started to get really interesting the next day, when the team began the tricky maneuver of removing the old Wiffpic and replacing it with the new version, which had its phenomenally delicate and precisely sited mirror spiking out of its very tip. Nothing untoward happened, either to it or to the camera, and the whole assembly slid into place with well-oiled ultraprecision, every part of it linking tightly with the bends and turns of the cavity inside which its predecessor had been living for the previous four years.

Nor was there any major problem with the culminating moment of the mission: the removal of the enormous High Speed Photometer and its replacement by the identically sized but wholly differently purposed COSTAR mechanism. Perkin-Elmer, shamed by its incompetence, had had no hand in the construction of this new mess of corrective optic mirrors. An entirely new company, called Ball Aerospace (descended from the company famous for making jam-preserving jars), out of Colorado, had won the loyalty and trust of NASA and been awarded the contract instead. And Ball had done well, with all measurements good, all fits exemplary, all tolerances met and matched. It took less than an hour to install the new optics package—it was almost anticlimactic, so flawlessly did it take place—before the crew members spent a final day tidying up and making cosmetic adjustments to their handiwork before leaving Hubble ready for business once again.

As a final dealing with Hubble, they opened its aperture door (the garbage bin lid, as it were), at the front of the telescope, and then attached their robotic arm to the monster spacecraft, lifted it very carefully out of the *Endeavour* cargo hold, and placed it gently beside (but now *outside*) their hull. Then, and as Captain Cook's crew would have said, they cast off springs and released the lashings that bound telescope and orbiter together. Finally—but this time as Cook's crew could not ever have imagined—they fired their thruster motors for a brief orbit-killing moment and headed back down to land.

Now Hubble, traveling still at seventeen thousand

miles per hour, but with its orbit very slightly and deliberately enhanced, reverted to its lonesome and unattended state, on its near-endless silver journey around the globe.

Had the repairs gone well? Was Hubble going to work? Was the humiliation over, and could the true value of this extraordinary device be seen, at last, for what it had always been intended to be?

All eyes now turned back to the control rooms, at the Mission Operations Center at Goddard, where they would resume flying the telescope; and more crucially, at the Science Operations Center at the Space Telescope Science Institute at Johns Hopkins, in Baltimore, where the astronomers would download and then translate the new images, and would immediately know their fate.

The long-ago inaugural glimpse from the space telescope, the basic concept for which had been advanced as long ago as the 1940s, had been termed First Light. The Great Disappointment, they might have termed it instead, the moment Eric Chaisson, also in Baltimore, examined the initial images and felt, as he later put it, that infamous total deflation in his gut.

Now this was December 18, 1993, some thirteen hundred days later. Back in 1990, First Light had been summertime. Now, for what was being called Second Light, it was winter nighttime. It was dark in Baltimore, it was silent, and it was cold. At the Science Operations Center, an astronomer ordered the tiny onboard electric motors to spin out the corrective mirrors inside CO-STAR and set them into their precisely allocated positions, to begin reordering the light beams inside Hubble. They also opened the shutter on Wiffpic, which had its own cleverly arranged optical correctors, buried deep within itself. Goddard obligingly pointed the enormous telescope toward a possibly fruitful portion of the sky. Everyone waited as the images slowly started to unscroll from the top to the bottom of their monitors.

Ed Weiler was there, the NASA engineer who had taken that first grim telephone call. Like everyone else in the room, he had his eyes locked on the screen. The next three seconds were the longest three seconds, Weiler said later, he had ever experienced in his life.

There was a sudden explosion of exuberance, applause, delight, and joy. The image on the screen was now complete, and it showed before everyone a vivid mass of stars, all in focus, with one star in the dead center occupying just a single pixel of the screen. One star, one pixel.

The image was sharp, perfectly, precisely sharp. No more fuzziness. No marshmallow. No soft edge. All was exact, aligned, impeccable, just as hoped for back when the project was a mere notion in a group of astronomers' heads. No other optical telescope that had ever been made and established on planet Earth (even those at the summits of great mountains in Hawaii, in Chile, on the Canary Islands, and in other places where the air was at its thinnest and clearest) could ever rival this.

Because down there *was* air—even when thin, it was heavy, windy, polluted, dancing with molecules, potent with distortion. Yet up here, nearly four hundred miles up, high above the troposphere, the stratosphere, the mesosphere, in what is now called the exosphere, where there was just the occasional hydrogen molecule drifting through, there was no air, and no distortion—and where now, at last, and thanks to the cleverness and cost of a whole set of new optics, humankind had a clear-eyed new viewing platform, like no other ever before, from which to observe.

Half a century after it was first conceived, twenty years after it was first designed, fourteen years after a computer in Danbury told the first polishing arm to travel across the great tablet of Corning quartz and begin to grind and shape its surface, and thirteen hundred and some days after that overflattened eight-foot mirror took its first long gulp of light from the universe wrapped around it, a repaired telescope with new precision optics was able to see clearly deep into the distance, and into the distant past of the cosmos.

The rest of the Hubble story is still being told today. Four further servicing missions have been up to it, as were scheduled long ago, each one tasked with breathing new life into what has become a beloved old silver workhorse, the greatest of NASA's great observatories. The longevity of the now almost venerable, if still no prettier, bird is greater than ever anticipated, and it is now expected to continue flying at least until 2030, maybe for a decade longer. It is by all accounts the most successful scientific experiment of modern times, maybe even of all times. And the images it has sent back, tens of thousands of them, have captivated all who see them. The eight-foot mirror, imperfect though it may be, has captured a vision of wonder and rapture to scientists and the lay alike, bringing the universe vividly to life.

Chapter 8

(Tolerance: 0.000 000 000 000 000 000 01)

Where Am I, and What Is the Time?

Each after each, from all the towers of Oxford, clocks struck the quarter-chime, in a tumbling cascade of friendly disagreement.

—DOROTHY L. SAYERS, *Gaudy Night* (1935)

Time is the longest distance between two places.

—TENNESSEE WILLIAMS, *The Glass Menagerie* (1944)

The offshore oil rig *Orion*, nine thousand tons of ungainly ironmongery being hauled slowly across the North Sea by a pair of tugs, was looking for a place to settle herself down and drill.

I was on the bridge of the lead tug, a small but exceptionally powerful craft called the *Trailblazer*, from Holland. *Orion*, her four jacked-up legs towering high over the drilling derrick itself and swaying in a dangerous-looking manner on the swells, had just completed a successful natural gas well five or so miles away. Now we were towing her to a place that the geophysicists back in Chicago had chosen, as the undersea geology looked promising for a new attempt.

It was March 1967, a bitter-cold early-springtime day, with a stiff nor'easterly breeze. I had worked on this rig for just one month. I was not yet twenty-three years old. The rig was worth ten million dollars, and Amoco Petroleum was renting her for eight thousand dollars an hour. Putting her down in the right place, exactly, was now, quite ludicrously, all down to me.

I had been given precious little by way of either a briefing or equipment to make sure that *Orion* settled herself properly. I had a two-way radio that let me talk to the tool pusher up on the rig. I had the British Admiralty maritime chart number 1408 (Harwich to Rotterdam and Cromer to Terschelling), which covered this portion of the North Sea. I had a confidential large-scale geophysical chart of the local ocean floor, fashioned by the American undersea survey teams, and on which someone had marked a big red X, as the place where the planners in Chicago now wanted the rig to put down. Written in pencil beside the X were the rig's coordinates, something in the order of 53°20'45" N, 3°30'45" E, but with the seconds of arc written to one or maybe even two decimal places.

Crucially, the tug's master also had a special chart overlaid with the curved lines (colored in red, green, purple) of the then-most-advanced radio navigation system known. This was the Decca Navigator chart, and like most captains of coastal-going vessels of the time, ours used it in conjunction with a large receiver that was mounted on a swivel at head height. This receiver, rented from the Decca company, sported four dials, three of them with what looked like clock hands, painted with luminous paint so that they could be read at night.

The receiver picked up the powerful radio signals that were being continuously transmitted from coastal radio stations, masters and slaves, that Decca had built on headlands and cliff tops on the English and German coasts of the North Sea. The signals, which were invariably short pulses, would go out from the master station, and then, a short moment later, the same pulse would be repeated by each of the slave stations. The delay between the master and slave pulses' reception by a receiver would vary depending on how far the receiver was from each of the slaves—and this, in turn, would allow the primitive computer in the receiver to deduce and determine, by taking a fix from the different distances to the various slaves, where on the chart the receiver was. The dials on the receiver would then show how far along which of the various lines of position (red, green, and purple) our little tug was proceeding. And because it showed where we were on three separate intersecting lines, we could draw them out from the Decca chart and—lo!—they would show where on the navigation chart, or on the geophysical chart, we actually were, to within an accuracy, the Decca makers insisted, of about six hundred feet.

What I had been told was that when I decided that the rig was exactly above the designated spot—and I knew from the Dutch skipper that there was a surface current running to the northwest, at about six knots, so I had to make allowances for that, as it would set the rig drifting a few score feet while it was settling—I had to instruct the tool pusher by radio to "Drop the legs!" He would immediately order the release of four sets of bolts, and the tall iron legs that were now towering above us would instantly plummet downward with four gigantic splashes, and would hurtle unstoppably down onto the seabed two hundred feet below. There they would pin themselves, by skewering themselves into the soft upper layers and thereby, with the addition of a set of anchors sent down later, fix the rig solidly into position for the next many weeks of her exploration.

We crept closer and closer. The fathometer pinged every few seconds, the depth below our keel showing a steady thirty-two fathoms. The Permian dome, which was to me just a vague pattern of half-inscribed lines on the geophysical chart that had been interpreted by specialists in Chicago to *be* a dome, crept nearer and nearer. For a few moments it appeared to be directly underneath where Decca told me the rig was, and I nervously fingered the Transmit button on the radio microphone, pressed it, looked up at the drilling platform, and spoke loudly into the microphone. In a tone as stern and as formal as a nearly twenty-three-year-old could muster, I commanded, "Drop the legs!"

An instant later, I saw the four small gusts of reddish rust smoke, and the enormous towers of tube iron trelliswork appeared immediately to collapse into themselves, to vanish quickly from sight. There was a fearful noise of screeching metal and a huge froth of roiling water. We ordered the seamen aboard the rig to release our towrope, and the same for the tug astern. The two tugs turned away and headed out to sea, away from the din, and then we stood off a mile or so and watched as the rig master ordered the jacking-up procedure to begin—noisily, once again, like the sound of a construction site jackhammer, and foot by foot, so the rig climbed up its own now stably rooted legs, pulling itself up by its own bootstraps, up and up, until it was a good forty feet above the waves, and then clear of most effects of storm and swell and surge below. Then someone aboard stopped the machinery, and silence fell, aside from the steady low howl of the gathering wind and the poundings of the swell.

The tool pusher came on the radio. He had just seen the bathymetry reports. "*All looks good,*" he said. "*The current kicked us off a little, maybe. We're about two hundred feet off the ideal. Pretty good for a beginner. Chicago will be okay with that. It's good enough. Go get some sleep.*"

They spudded the well later that evening, and then drilled night and day for the next three weeks. We hit gas at six thousand feet, a good and powerful flow of what back in the sixties were the blessings of raw hydrocarbons. A week later we capped the well off, leaving it to be connected to a producing field by a later gang of workers, and *Orion* and her crew departed with another couple of heavy-haul tugs for further hunting grounds in the sea.

In due course, I left the rig, then the company, and eventually the profession of petroleum geologist altogether, but the knowledge that I had once helped locate a nine-thousand-ton drilling rig over a Permian salt dome in the heaving middle of an ocean, and had managed to do so with sufficient accuracy to create a flowing gas well, stayed with me for many years.

We had reached to within two hundred feet of the mark, a figure that seemed to me at the time a very considerable achievement. But being two hundred feet off an X drawn on a chart is, by today's standards, unimaginably imprecise, a total fail. Places on the surface of the planet can now be located within centimeters (millimeters soon), and they can be so located because of the making of a technology that would eventually replace Decca and LORAN and Geo and Transit and Mosaic and all the other proprietary and radio-based navigation systems of the time, and would indeed also replace the sextant* and the compass and the chronometer

---

* Before Decca and LORAN and long before GPS, the combined use of a sextant and a good chronometer enabled a skilled mariner to find his position at sea with a fair degree of accuracy. As a very unskilled sailor of a small schooner in the Indian Ocean in 1985, and under the supervision of a practiced Australian skipper, I managed, by using just these tools, together with a good set of charts and a ready-reckoning log towed astern, to

and all the various navigation bridge furnishings with which sailors had been determining their positions for centuries.

It's called GPS.

The basic principle of this new technology was unexpectedly born of the development of quite another.

It was in Baltimore, on Monday, October 7, 1957, when two young scientists, William Guier and George Weiffenbach, arrived at their Applied Physics Laboratory at Johns Hopkins University, enthralled like all American scientists by the fact that, for the first time ever, an artificial moon was currently in orbit around Earth.

It was *Sputnik*, a two-hundred-pound, twenty-inch-diameter sphere of polished titanium alloy that the Soviet Union, to the chagrin of the American public, had launched the previous Friday, and which was now orbiting Earth once every ninety-six minutes. The *New*

---

sail unaided the 1,300 miles from Diego Garcia to Mauritius. My daily on-voyage accuracy was seldom better than a couple of miles, but finally seeing the four white flashes of the Flat Island light off the port bow late one night, and realizing we were just a few miles north of Mauritius itself after ten days' sailing across an empty ocean, also remains for me, just as with the oil rig of twenty years before, a powerful memory of navigational success.

Many years of occasionally partisan bickering led eventually to the acceptance of Vermont-born Roger Easton (third from left) as the inventor of GPS, while he was working at the U.S. Naval Research Laboratory in Washington, DC.

York Times Sunday edition had reported (on page 193 of its *360-page* paper) that the device was continuously emitting radio signals from a tiny transmitter on board. Guier and Weiffenbach (both computer experts, their most recent work being on hydrogen bomb simulations and microwave spectroscopy, respectively) reckoned they could probably determine exactly where the satellite was by recording and then analyzing its radio signals.

Accordingly, they used the specialized radio receivers in the lab to tune in to *Sputnik*'s frequency, and listened intently to the regular heartbeat of its transmissions (a high-pitched beep, sent out a little faster than twice a second) and recorded it on a high-fidelity tape deck. They then analyzed the frequency of the signal and, as they suspected might be the case, heard it alter very slightly as the satellite rose above the horizon, as it then passed directly overhead of them in their Baltimore lab, and finally then set down once again. The frequency change they observed was the Doppler effect—the classic example being the change of perceived frequency of the horn of a passing train—and for the first time ever, it was shown by this pair of physicists to be both detectable and measurable in a satellite signal.

Shortly thereafter, and by employing as powerful a computer as was available—the Applied Physics Laboratory had a brand-new Remington UNIVAC at hand—the pair was able to digitize the signal and, from the varying frequencies that had now been converted into numbers, to calculate with fair precision how far away *Sputnik* was on each one of its orbits. The frequency when the satellite was directly above them was the true frequency of the signal; from the variations, the first when it was approaching them and then again when it was moving away from them, gave them the basis for a calculation (as they knew from its circumorbital time that it was moving around the planet at about eighteen thousand miles per hour) of how far away it was.

Their sums (which they then also applied successfully to predicting the orbits of *Explorer I*, once America had entered the space race) involved many weeks of computer time, and would have profound consequences. For the following March, the chairman of the Applied Physics Laboratory, Frank McClure, realized that his two young colleagues had unwittingly stumbled upon the makings of an application that could have worldwide use.

As he told the pair when he hauled them into his office and demanded that they close the door, if an observer on the ground could establish with precision the position of a satellite in space, then the opposite, the numerical reciprocal, could be true as well. From the position of the satellite, one could compute the exact position back on Earth of the person or machine that observed it.

Guier and Weiffenbach had never noticed what in retrospect was blindingly obvious, nor did they immediately appreciate the corollary: that a satellite navigation system based on this simple Doppler principle could do for ships and trucks and trains and even for ordinary civilians, mobile or stationary, what the sextant, the compass, and the chronometer had done for centuries past for mariners, and what LORAN and Decca and Gee were doing at that very moment. It could tell them where they were; moreover, it could tell them what direction they should take if they wanted to go somewhere else. "It occurred to me," wrote McClure, in a famous memo that claimed the prize for Guier and Weiffenbach, "that their work provides a basis for a relatively simple and perhaps quite accurate navigation system."

Quite accurate indeed: the U.S. Navy, which paid for much of the APL work in Baltimore, did some back-of-envelope calculations and came up with the notion that with a good number of satellites, the location of someone's or something's position (that of a ship or a submarine) could be achieved within perhaps a half mile. And while that may not be as precise as the six hundred feet guaranteed to Decca, there was a further significant difference, an advantage that was especially relevant in these times of the gathering problems that related to the Cold War. The radio-based Decca-like systems then employed by ships, and by oil rig location tugs such as *Trailblazer*, were hardly secure, as their transmitters were all based on land, and could easily be put out of service by a canny foe. A system that involved satellites out in space, however, was by its very nature much more protected from outside tampering and interference, from surveillance and from sabotage. Moscow, the enemy du jour, would find it difficult to mess with it or find out anything from its use.

The U.S. Navy, at the time, was looking for a fool-

One of the early Transit-system satellites. Launched for the U.S. Navy in the 1950s and '60s, it used Doppler-based navigation to establish to within three hundred feet the position of American strategic submarines. Transit is seen as the first operational use of a system that eventually led to the modern Global Positioning System, GPS.

proof, secure, and accurate means of locating its fleet of Polaris-armed nuclear submarines, and thus was born the Doppler satellite navigation system known as Transit. A prototype satellite was successfully put into orbit in 1960, and no more than six years after McClure's memo (seven years after the launch of *Sputnik*), a flotilla of U.S. Navy Transit satellites was in orbit around Earth, and the first true satellite navigation system was declared to be fully operational.

Fifteen satellites were built, rather inelegant-looking and insect-like creatures with four solar-panel wings and a long torso attached to a transmitter that acted as a boom to keep the antennas pointing down to Earth. At least three devices at a time were kept up in polar orbit, six hundred miles high. As the world turned beneath them, they swept the land and sea masses below, rising and setting like the sun and sending out signals to receivers on the ground that would be Doppler-affected as they moved toward, over, and away from the receivers. Earth stations, equipped with enormous computers with tape drums whirling back and forth, would predict the true orbit of each satellite as it appeared in the sky, and would radio these data to the ships and submarines that needed to know where in the world they were. And cumbersome and slow though it might have been, and available only once every few hours in its early days, the system most certainly did allow U.S. Navy ships anywhere in the world, at any time of the day or night, and in any weather, to learn their more-or-less-exact location.

Within fifteen minutes of tracking a passing satellite, a ship could know where it was to an accuracy of three hundred feet. And the Polaris-carrying submarines of the strategic fleet, which were privileged to use an enhanced and highly secret version of the software (i.e., of the signal giving the satellites' correct orbits), were said to be able to tell their position to within sixty feet. It was clearly a far more robust* system than Decca or LORAN or its other radio-based competitors, and it endured: the Transit system was in use until 1996, for more than thirty years. It was made available to commercial ships in 1967, and at its height, as many as eighty thousand non-navy vessels were using the system, "the largest step in navigation since the development of the shipboard chronometer," as a program manager put it.

The world was moving faster, nuclear weapons were ever more dangerous, enemies were wilier, critical infrastructure was more demanding—and figures of what the

---

* However robust the system may have been, it did not entirely survive some very poor planning and decision making by a sister agency of the U.S. government, the Atomic Energy Commission. The navy's Transit 4B satellite was launched in June 1961 and was sashaying quietly along its planned orbit, sending out its signals with impeccable regularity. However, a little more than a year later, the AEC launched a rocket with a powerful hydrogen bomb in its nose cone, which exploded as planned four hundred miles above Earth, near Hawaii. The AEC had forgotten to check, and it blew the poor little Transit out of the sky, one of several orbiting bodies that were damaged or destroyed that summer night. It also knocked out streetlights in Honolulu. Only the New Zealand Air Force was pleased, as the explosion lit up the South Pacific for a sufficient time to allow aircraft on exercise to find their target submarines. The exercise planners later claimed this was cheating.

navy was calling "pinpoint accuracy" (e.g., six hundred feet, three hundred feet, two hundred feet, sixty feet) were clearly pinpoint in name alone. Moreover, fixes were available only once an hour, and they took as much as fifteen minutes to evaluate. Also, the procedure required ground stations and faraway banks of computers and small armies of navy personnel, each one vulnerable to human error no matter how good his or her training.

The new world order demanded something better, quicker, more reliable, much more secure, and very much more precise. Doppler shift–based navigation, good and reliable though it was, when confronted by the technical realities of the newer, faster, more threatening environment, clearly couldn't cut it. Then, in 1973, a Vermont country doctor's son, Roger Easton, came up with something that very clearly could. It involved the question of time, and of the clocks that record its passage. Indeed, the physical principle involved is known as passive ranging, and in its essence, it is disconcertingly simple.

Suppose there are two clocks that are entirely reliable and show exactly the same time. Suppose further that one clock is in London, the other in Detroit, and that both clocks are linked by a video stream, are both on Skype, or FaceTime, or WhatsApp. In this scenario we have total faith in the exactness and accuracy of the two timekeepers, and we know with total certainty that they were both set at the same time, that both are consequently displaying the same time.

And this is certainly true for those observers, those clock watchers, who are in the same rooms as each of the two clocks. But for the observer in London, who is looking at the displayed image of the clock in Detroit that is coming across to his screen, there is actually a slight, very tiny difference. To him, it appears that the Detroit clock is the tiniest fraction of a second (almost exactly one-fiftieth of a second, in fact) late compared to the clock beside which he is sitting in London. He knows for certain, though, that both clocks are actually showing the same time. He knows also that the speed of the signal between them, the speed of light, is a constant. So the discrepancy must therefore be the result of the only unknown variable in this scenario—and that, clearly, is the distance between Detroit and London over which the signal has to travel.

Roger Easton, who at the time worked for the U.S. Navy's then-named Space Applications Branch in the Rio Grande Valley of South Texas, and who created the infamous "space fence," a vast array of detectors claimed to be able to map any satellite passing over U.S. territory, realized that the simple fact of the perceived difference in the clocks' times offered up a valuable piece of information. It gave him a number from which (because light travels at a certain fixed absolute velocity) he could calculate the distance between the two cities. In one second, light travels 186,000 miles. In one-fiftieth of a second (the measured delay in this example), it will have traveled 3,700 miles. So the distance between Detroit and London, according to this time-based calculation, is 3,700 miles—which is, essentially, what it turns out to be.

So Easton promptly devised a simple experiment, and invited senior navy officer colleagues to watch. But for this he didn't use clocks: back in the mid-1960s, very precise atomic clocks, though they had already been invented (and will be described shortly), were far too bulky to employ in the experiment he had in mind. Instead, he employed a quartz oscillator, but with a costly and complex (but conveniently small) device known as a hydrogen maser, which would give a wholly reliable and exactly constant frequency standard.

He made two such devices. One of them he put in the trunk of a convertible car that was owned by an engineer friend named Matt Maloof; the other he kept at the naval station in which he was working in South Texas. While the observers were watching the oscilloscope screens he had hooked up in the lab, he ordered Maloof to drive the car as far and as fast as possible down a road,

Texas Route 295, which was unfinished at the time, and thus empty. All the while as he sped away, his transmitter was busily sending out signals that were being received back at HQ by an oscillator that was set to exactly the same frequency as the transmitter.

As the distance between the car and the office increased, so did the discrepancy between the two numbers, and it did so solely because of the distance, as all else (the frequencies of the two devices and the speed of signal transmission, the speed of light) was constant. The navy officers watched, fascinated. As the calculations came in, more or less instantly, they could tell exactly how far away Maloof's car was, how fast he was going, and when he changed direction. They noted with particular admiration and frank astonishment as the number changed noticeably at the one point when Maloof, now driving scores of miles away, changed lanes. The demonstration was a consummate success: in principle, clock-difference navigation systems were shown to work, and far more easily than anyone had imagined.

The navy promptly released funds for further research—a trivial amount, and not enough for the launch of a satellite to test the idea in what the military likes to call the real-world environment. Meanwhile, still other ways of determining position were being thrown up by laboratories across the United States—the notion that this was a duel to the death between Doppler-based systems and clock-based systems took some while to be distilled from a mess of conflicting technologies, and personalities, and branches of the disciplined services. There is to this day much unfriendly rivalry between supporters of the navy's Roger Easton and those of an air force combat-hardened officer named Bradford Parkinson,* who some like to think fathered the system. There is still dark talk of a "GPS Mafia," and occasionally even today one reads ill-tempered writings by supporters of the two claimants. Eventually, though, the clock-based system won out, and in 1973, the U.S. Air Force, having won part of the battle by prising op-

---

* Brad Parkinson's air force career was deeply involved in the "automated battlefield" idea, with a special interest in the formidably armed AC-130 aircraft, which has a reputation for being the "terminator," the ne plus ultra of fixed-wing gunships. Parkinson's association with GPS derives mainly from a legendary meeting, the so-called Lonely Halls Meeting, held in an otherwise near-deserted Pentagon over the Labor Day weekend of 1973, when the outlines of the GPS architecture were discussed by a handpicked group of air force officers. Parkinson saw the importance of GPS as allowing aircraft "to drop five bombs in the same hole." Roger Easton, by contrast, liked to think of his work as the poetic continuation of John Harrison's timekeeping obsession of two centuries before, though linking time and space with modern technology.

Easton's rival claimant as inventor of GPS was the U.S. Air Force colonel Bradford Parkinson, otherwise well known for his work on the so-called automated battlefield. His vision for GPS was very much a military one, while Easton, more poetically, thought of the system as a natural successor to John Harrison's eighteenth-century work on both longitude and highly accurate clocks.

erational control from the plan's originators in the navy, began the construction of the satellite system that would be at the core of what would be called the Navstar Global Positioning System—later to be simplified to what it now familiarly goes by, GPS. And to Roger Easton went the laurels: he was in due course awarded the National Medal of Technology and a slew of other distinctions, including induction into the National Inventors Hall of Fame for being the system's principal inventor.

There were technical problems aplenty for the proposed system, and so the constellation of satellites needed for the worldwide coverage was sent up in series (or blocks, as they were called), to work out the kinks. The first ten devices of Block 1 were placed into orbit between 1978 and 1985, with GPS as a working system being formally inaugurated in February 1978, though for the exclusive use initially of the U.S. military. Some military strikes (on Libya's leadership, for example) were then carried out with the use of GPS targeting. Weapons were designed and bombs were fitted with inbuilt GPS—smart bombs, as they were known. Subsequently, entire wars (the Gulf War of 1991 being arguably the first) were fought with GPS as an essential part of planning and tactics. (The lead tanks that headed the columns of troops into Kuwait were all equipped with GPS receivers.) There have since been seventy GPS satellites put into medium Earth orbit, about twelve thousand miles up. Thirty-one remain, all made either by Lockheed Martin or Boeing, most launched by the U.S. Air Force using Atlas V rockets, most sent up from Cape Canaveral, most sent up since 1997—so some of the satellites are quite geriatric. Together they provide the operational backbone of a system that is now regarded as essential to all, a common good, and offered by the U.S. government wholly free of charge.

A truly common good for the simple reason that GPS, though owned by the U.S. government, is now a system fully available to civilians, with almost no restrictions. Initially it was top-secret, a component of the nuclear strategic arsenal designed to make certain that planes carrying atomic bombs and submarines armed with nuclear-tipped missiles always knew where they were to a high degree of accuracy, and that their weapons knew their targets' locations to within margins of just a few meters. Then, in the aftermath of the shooting down in 1983 of Korean Air Lines Flight 007 by Soviet fighters after it accidentally strayed into forbidden airspace over Sakhalin Island while flying from Anchorage to Seoul, Ronald Reagan decided that civil users (airlines initially, and then ordinary civilians, too) should have equal access to the technology. To withhold deliberately a means of accurately determining one's location was considered morally questionable, Reagan's White House decided, even when ranged against the strategic advantage of keeping the information to oneself, as was claimed by the military. Besides, the Soviet Union was then on the brink of collapse, and was busily engaged in making its own global navigation system. (That system now exists, and is called GLONASS. There is also a pan-European system, called Galileo; and a Chinese system, Beidou, is up and running and will presumably soon become as ubiquitous as GPS.) For now, though, GPS itself remains paramount, and it has to be assumed, as long as no malicious hackers manage to penetrate American defenses, that it will remain supreme for some years to come.

For many years after the freeing of GPS for civilian use, the still-skittish U.S. Defense Department, fretting that the common man should not be privy to the exact whereabouts of the Oval Office, certainly not to the nearest meter or two, demanded that the air force introduce a deliberate error into the system, corrupting it slightly so that civilian users could never know a location to a better accuracy than one hundred fifty feet horizontally and three hundred feet vertically. Yet that restriction, what was called selective availability, was scrapped in 2000 on the orders of President Clinton. Ever since then, users worldwide have been able to use GPS receivers in everything from their cars to their telephones to wristwatches to handheld devices taken on hunting expeditions and weekend sailing vacations, to get accuracies of just the barest few meters. Survey teams, using special receivers and being able to wait while more and more satellites swim into view—at least four satellites must be in line of sight to give a decent reading; some surveyors wait until they can communicate with as many as twelve—claim to be able to site with a precision of just a few millimeters.

The whole system is currently run from the tightly guarded Schriever Air Force Base, on the dusty east-sloping plains that spread out in the rain shadow of the Rockies near Colorado Springs—close to the famously immense bunker under Cheyenne Mountain from where the United States is supposedly protected from nuclear attack. Schriever looks after almost all the Defense Department's hundreds of satellites, most of which are intelligence-gathering and highly secret, and which fly or hover overhead, bent on all manner of dubious tasks. Buried deep within the air force bureaucracy, though, and buried equally deep behind

Schriever Air Force Base, in the rain-shadow flatlands of Colorado, is where GPS, an American Defense Department–owned system, is managed and controlled, under conditions of formidable security.

layers of protection within the huge and highly secure complex of the base itself, are the men and women of the Second Space Operations Squadron, or 2 SOPS, whose duties, under the somewhat inevitable American motto "Pathways for Peace," are almost exclusively devoted to managing and maintaining the constellation of thirty-one satellites that make up America's GPS. The Master Control Station here checks the health of every satellite as it appears above the horizon, and a network of sixteen monitoring stations around the world ensures that, at any one time, at least three sets of eyes, assisted by banks of electronic enginework and hyperfast computational power, are supervising each of the satellites at all times, night and day.

Four of these stations have complex antennas that can beam information up to the satellites—information that includes, and crucially, corrections measured in millionths of seconds of the atomic clocks that each of the satellites carries on board. For, while the fact that each satellite sends out its precise position information is important, the fact that it is also sending out a super-accurate time signal is of truly extraordinary importance, as the function of the GPS goes some way beyond simply assisting the planet with its navigational needs. GPS clocks, it can fairly be said, run most of the

A technician in the ops room of the U.S. Air Force Second Space Operations Squadron, which manages the thirty-one GPS satellites that offer to receivers across most of the world highly accurate navigation and position information.

modern world's economy, and ensure that it runs on time, and to within the tiniest fractions of a millisecond.

In summary: the complex utility of the flotilla of GPS satellites hovering or scooting above Earth is about time. The signal's so-called time of transmission is a number instantly compared to its "time of arrival," the immediately calculated difference being the "time of flight"—and from four times of flight, there can be computed (by dividing the numbers by the speed of light) four distances, and from the triangulation of those four distances can be derived the receiver's exact position, to within five meters, it is generally said—except that, as the clocks get better and better, and all the calculations are based on ever-more-precise calculations of time, the accuracies of locations will get better and better, too. In terms of basic geometry, America's GPS and its sister systems in Russia and China and Europe operate with elegant simplicity, but at the heart of each of them are devices of immense sophistication in terms of the accuracy of their offerings, which leads to quite astonishing degrees of precision in the tasks for which GPS is currently employed.

And those tasks go far beyond guiding a ship safely into harbor, or taking a motorcar through the streets of Ulaanbaatar at rush hour. Cellular telephony, agriculture, archaeology, tectonics, disaster relief, mapping, robotics, astronomy—almost any human activity that requires knowledge of time and place is improved with the ever-greater precision of the information that acts as guide.*

---

* Many of the major achievements of nineteenth-century cartography, when checked against modern GPS-derived data, have turned out to be surprisingly accurate. The thirteen-year-long Great Trigonometric Survey of India, begun by Sir George Everest in 1830, employed thousands of men working in glaciers, jungles, swamps, and hot deserts with iron chains and theodolites, and used as its baseline the fourteen-hundred-mile "Great Arc" between the Himalayas and Cape Comorin. A 2003 resurvey of the arc using laser technology and satellites showed the Victorian survey to be out by only 0.09 percent. Moreover,

Or so we are supposed to believe. Philosophically, morally, psychologically, intellectually, and—dare one say it—spiritually, there are troubling aspects to human-kind's ever-greater reliance on devices and techniques of ever-enhanced precision. The same doubts that were raised by the machine breakers of the seventeenth century, by those who later mourned the passage of crafts-manship or who today react with deer-in-the-headlights bewilderment at the invisible magic of electronics, remain. (I shall return later to the question of the perceived and the actual benefits of precision.)

In personal terms, one thing, however, is clear. Half a century on, it still rankles that I put that oil rig down in the ocean two hundred feet off its target. Yes, it drilled, it hit gas, it was a success, but that two hundred feet—that distance bothers me every time I think about it. It was inaccurate. It was imprecise. If only,

---

the vertical surveying of India was just as good as the horizontal: the team calculated the height of the Himalayas' highest mountain—Peak XV—as 29,002 feet; subsequent surveys have indicated 29,029 feet. The peak was later to be named in English Mount Everest—though pronounced *Evv-rest*, unlike Sir George, whose family name was *Eve*-rest, the first syllable stressed. The peak stands as a potent reminder of the quality of Victorian measuring precision.

I say to myself these days, if only I had had access to GPS, to a technology that was already being discussed by the team of physicists in Baltimore assessing the consequences of the launch of *Sputnik*. Then I could have put that rig down to within ten feet or better, and all would have been content. Yet, even though back in Baltimore they had been talking about satellite navigation for the previous decade, and even though the first steps to build a system had been taken, it would be another twenty years before a constellation of useful satellites was launched, and before I and thousands like me had the tools to allow us to do better than we were doing.

And, in any case, would ten feet, in practical terms, truly have been much better than two hundred? After all, as the tool pusher said, two hundred feet was "good enough."

I have a Japanese friend who works as a navigation officer on a deep-ocean research vessel in some of the most distant quarters of the northwest Pacific Ocean. On the bridge, he has a GPS annunciator that communicates with twelve of the GPS satellites—most iPhones talk to three or four—and, as a result, is able to know his position on a trackless sea to within just a couple of centimeters. Not a couple of yards, not a couple of meters, nor even a couple of feet, but a couple of *cen-*

timeters, and that out in the swell and loneliness of the middle of an ocean.

I remembered well the Amoco tool pusher allowing that a two-hundred-foot error at sea was good enough. When I told my Japanese friend of the sanguine attitude of the men on the rig, he laughed. Of course, he said, that was back in the sixties. But that is not what precision is about, he said. "Good enough" is absolutely not good enough.

There will come a time, he then added, with his voice rising, when centimeters are simply not good enough, either, when we'll need to know where we are at sea to within just millimeters. "There are no limits to precision, no end to the need for absolute perfection."

His words echo still, like the mantra of a new religion. Or of a cult.

Chapter 9

(Tolerance: 0.000 000 000 000 000 000 000 000 000 000 000 01)

Squeezing Beyond Boundaries

One can never know with perfect accuracy both of those two important factors which determine the movement of one of the smallest particles—its position and its velocity. It is impossible to determine accurately both the position and the direction and speed of a particle at the same instant.

—WERNER HEISENBERG, *Die Physik der Atomkerne* (1949)

O nce every few weeks, beginning in the summer of 2018, a trio of large Boeing freighter aircraft, most often converted and windowless 747s of the Dutch airline KLM, takes off from Schiphol airport outside Amsterdam, with a precious cargo bound eventually for the city of Chandler, a western desert exurb of Phoenix, Arizona. The cargo is always the same, consisting of nine white boxes in each aircraft, each box taller than a man. To get these profoundly heavy containers from the airport in Phoenix to their destination, twenty miles away, requires a convoy of rather more than a dozen eighteen-wheeler trucks. On arrival and finally uncrated, the contents of all the boxes are bolted together to form one enormous 160-ton machine—a machine tool, in fact, a direct descendant of the machine tools invented and used by men such as Joseph Bramah and Henry Maudslay and Henry Royce and Henry Ford a century and more before.

Just like its cast-iron predecessors, this Dutch-made behemoth of a tool (fifteen of which compose the total order due to be sent to Chandler, each delivered as it is made) is *a machine that makes machines*. Yet, rather than making mechanical devices by the precise cutting of metal from metal, this gigantic device is designed for the manufacture of the tiniest of machines imaginable, all of which perform their work electronically, without any visible moving parts.

For here we come to the culmination of precision's

quarter-millennium evolutionary journey. Up until this moment, almost all the devices and creations that required a degree of precision in their making had been made of metal, and performed their various functions through physical movements of one kind or another. Pistons rose and fell; locks opened and closed; rifles fired; sewing machines secured pieces of fabric and created hems and selvedges; bicycles wobbled along lanes; cars ran along highways; ball bearings spun and whirled; trains snorted out of tunnels; aircraft flew through the skies; telescopes deployed; clocks ticked or hummed, and their hands moved ever forward, never back, one precise second at a time.

Then came the computer, then the personal computer, then the smartphone, then the previously unimaginable tools of today—and with this helter-skelter technological evolution came a time of translation, a time when the leading edge of precision passed itself out into the beyond, moving as if through an invisible gateway, from the purely mechanical and physical world and into an immobile and silent universe, one where electrons and protons and neutrons have replaced iron and oil and bearings and lubricants and trunnions and the paradigm-altering idea of interchangeable parts, and where, though the components might well glow with fierce lights or send out intense waves of heat, nothing moved one piece against another in mechanical fashion, no machine required that measured exactness be an essential attribute of every component piece. Precision had by now reached a degree of exactitude that would be of relevance and use only at the near-atomic level, and for devices that were now near-universally electronic and that obeyed different rules and could perform tasks hitherto never even considered.

The particular device sent out to perform such tasks in Arizona, and which, when fully assembled, is as big as a modest apartment, is known formally as an

It takes an enormous machine to allow for the making of something so infinitesimally tiny as a computer chip. This Twinscan NXE:3350B photolithography machine, made by the Dutch company ASML, would fill three jet cargo aircraft. Intel, the world's biggest chip maker, buys these $100 million machines by the score.

NXE:3350B EUV scanner. It is made by a generally unfamiliar but formidably important Dutch-registered company known simply by its initials, ASML. Each one of the machines in the order costs its customer about $100 million, making the total order worth around $1.5 billion.

The customer whose place of business is in Chandler—a conglomeration of huge and faceless buildings that are known in the argot as a "fab," or fabrication plant, for in line with this new world order, factories that make metal things are being supplemented by fabs* that make electronic things—could easily afford the sum. Intel Corporation, a fifty-year-old tentpole of the modern computer industry, has current assets well in excess of $100 billion. Its central business is the making, in the many fabs it has scattered around the planet—the one in Chandler is known as Fab 42—of electronic microprocessor chips, the operating brains of almost all the world's computers. The enormous ASML devices allow the firm to manufacture these chips, and to place transistors on them in huge numbers and to any almost unreal level of precision and minute scale that today's

---

* Or foundries, plucking from the seventeenth century a term that dealt in the crudities of ironworking to describe a twenty-first-century phenomenon wreathed in the delicacies of electronics.

computer industry, pressing for ever-speedier and more powerful computers, endlessly demands.

How the two tasks are managed, the making of the chips and the making of the machines that make the chips, are two of the more memorable and intertwined precision-related sagas of recent years. The technology that now binds the two companies together* is performed on such an infinitesimally minute scale, and to tolerances that would have seemed unimaginably absurd and well-nigh unachievable only decades ago, that it is taking precision into the world of the barely believable—except that it is a world that manifestly must be believed, a world from which, one can argue, modern humankind benefits mightily and for whose existence it is all the better, an assertion with which both Intel and ASML would readily agree.

Gordon Moore, one of the founders of Intel, is most probably the man to blame for this trend toward ultra-precision in the electronics world. He made an immense fortune by devising the means to make ever-smaller and smaller transistors and to cram millions, then billions of

---

* The mutual dependency of the two companies is such that, in 2012, Intel spent four billion dollars to acquire a 15 percent stake in ASML, trusting that the Dutch firm's researchers would use the funds to come up with ever-more-precise and economical devices for manufacturing microprocessor chips.

For now, the law advanced by Gordon Moore (seated) in 1965, when he ran Fairchild Semiconductor, in which he forecast that integrated circuit performance would double every year (a figure he later and prudently revised downward, to double every two years), still obtains, though most agree it is reaching the limits of possible performance.

them onto a single microprocessing chip, the heart and soul of every computing device now made. He is best known, however, for his forecast (made in 1965, when he was thirty-six years old and evidently a coming man) that from that moment onward, the size of critical electronic components would shrink by half and that computing speed and power would double, and would do so with metronomic regularity, every year.

An amended version of Moore's law, as a colleague promptly named the pronouncement, has since assumed the status of Holy Writ, not least because it proved to be more or less correct, its forecasts uncannily accurate.

Yet, as Gordon Moore himself has noted, his law has served not so much to *describe* the development of the computer industry as to *drive* it. For firms that make computer chips seem nowadays to be bent on manufacturing them to the minutest and most ever-diminishing tolerances, just in order to keep the Moore's law bandwagon rolling.

The electronics journals of recent years have been filled with essays showing how this new chip, or that new processor, or that newly designed motherboard, offers a further indication that Moore's law is still in effect thirty, forty, fifty years after it was first promulgated. It is rather as though Moore, if unwittingly, has become some sort of venerably wise Pied Piper, leading the industry to go ever faster, to make devices ever smaller and ever more powerful, just to fulfill his forecast. And to do so even though, quite possibly, many consumers, heretical and Luddite-inspired though the thought may be, find it quite unnecessary. They might rather wish for some settling, for a time of calm, a moment of contentment, rather than to be gripped by the perceived need to buy the latest iPhone or the machine equipped with that newest and fastest microprocessor (though not entirely certain what a microprocessor is, or does), to ensure that everyone is Keeping Up with Moore.

The numbers are beyond incredible. There are now

more transistors at work on this planet (some 15 quintillion, or 15,000,000,000,000,000,000) than there are leaves on all the trees in the world. In 2015, the four major chip-making firms were making 14 trillion transistors every single second. Also, the sizes of the individual transistors are well down into the atomic level.

Having said this—this very last assertion means that fundamental constants of physics may have their own plans for quieting matters down—I must point out that it is beginning to appear as though conventional electronics were about to reach some kind of physical limit, and that Moore's law, after five giddy decades of predictive accuracy, may be about to hit the stops. Not, of course, that this will inhibit the computer industry from creating some entirely new technology to take its place, as is clearly occurring right now. Whether Moore's law will continue to apply to that new technology, however, remains to be seen.

Gordon Moore was born in 1929, the son of the sheriff of San Mateo County in Northern California. The idea for the device that would dominate Moore's professional life, the transistor, had just been foreshadowed, and by a man Moore was never to meet. Julius Lilienfeld, who left Leipzig for Massachusetts in the 1920s, had drawn up a series of hesitant and untidy plans for making an all-electronic gateway, a device that would allow a low-voltage electrical current, by the employment of a substance then known as a semiconductor, to control a very-much-higher-current flow, and either to switch it on and off at will or to amplify it—all without moving parts or exorbitant cost.

Hitherto such work had been performed by a breakable, costly, and very hot (while working) glass tube–encased diode or, later, triode: the solid-state version of this, which Lilienfeld dreamed might one day be made, could replace it, and consequently make electronics cool, small, and cheap. He patented his idea in Canada in 1925, creating drawings of "A Method and Apparatus for Controlling Electric Currents." His scheme was entirely conceptual, however: no such device could be made with the technology and materials available at the time—all that existed was his idea and his newly announced principle.

Time passed; the idea endured. It took twenty years for Lilienfeld's concept to be made real, and a working transistor was indeed made and was well into development by the time the young Moore entered the California university system as a competent but not apparently overly gifted student in San Jose.

Two days before Christmas in 1947, the Bell Labs physicists John Bardeen, Walter Brattain, and William

Shockley—the last a difficult man, later to be reviled as a keen proponent of eugenics; his cool calculation of likely wartime casualties helped tip the scales in favor of President Truman's decision to drop the atomic bomb on Hiroshima and Nagasaki—had unveiled the first working device. They would win the 1956 Nobel Prize in Physics for doing so: in his lecture, Shockley remarked of what had been invented that "it seems likely that many inventions unforeseen at present will be made." He knew only the half of it.

The invention was not yet called a "transistor"—that term would enter the lexicon a year later, as it exhibited a blend of electrical properties that employed the words *transfer* and *resistor*. It was also far from being a small thing: its prototype is now preserved in a Bell

John Bardeen, William Shockley, and Walter Brattain (*left to right*), joint winners of the 1956 Nobel Prize in Physics for their discovery of the "transistor effect." Bardeen would win again, in 1972, for his work on semiconductors, becoming one of only four people to have won the Nobel twice.

Labs bell jar, and with its wires and various components and the crucial semiconducting sliver of a previously little-regarded silvery light-metal element, germanium, it occupies the volume of a small child's hand.

Yet, in a matter of months, the devices that would deploy the so-called transistor effect started to be made very much smaller, and by the time, in the mid-1950s, that the first transistor radios were on the market, the little glass thimble, with its characteristic three wires protruding from its nether region—one introducing the gate voltage into the transistor; the other two, unattractively called the source and the drain, being live only when voltage is applied via the gate—had become a familiar sight.

Small and miraculous in their abilities these glass-and-wire thimbles may have been, but they were still far from being tiny. Tiny became theoretically possible only after the invention of silicon wafer–based transistors in 1954, and then, most crucially, with the making in 1959 of the first entirely flat, or planar, devices. By this time, the young Gordon Moore had entered the picture, having been fully steeped in science during years spent at Berkeley, Caltech, and Johns Hopkins. He had now left the world of the academy to enter commerce, and to explore the commercial possibilities of the fledgling semiconductor industry. He did so specifically at

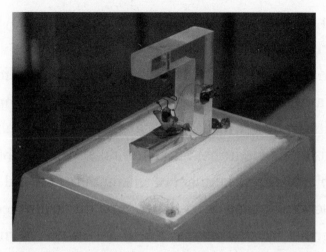

The first transistor, invented by Bell Labs in New Jersey shortly before Christmas 1947. Arguably no other twentieth-century invention has been so influential, and in the story of precision, its creation marked the moment when moving mechanics gave way to immobile electronics, when Newton passed the mantle to Einstein.

the behest of William Shockley, who had left Bell in 1956 and headed out west to Palo Alto, there to set up his own company, Shockley Transistors, and search for the first of his predicted "many inventions unforeseen."

His doing so marked the establishment, essentially, of what would become Silicon Valley, the then-unbuilt cathedral for the coming religion of semiconducting. For Shockley, who, thanks to his Nobel Prize and his reputation, was well financed enough to hire whomever he wanted, swiftly assembled a stable of scientific rarae aves, including Gordon Moore as his chief chemist, together with a cluster of equally bright young physicists and engineers.

Shockley promptly drove them all mad. Within a year, a group of eight of his first hires, all complaining bitterly about his tyrannical and secretive behavior and his overtly paranoid mien (and his inexplicable and unexplained abandonment of silicon as the central semi-conducting element of his firm's research), stormed out. The group, later to be known by Shockley's dismissive term for them, the Traitorous Eight, formed in 1957 a new company that was to change everything. Their start-up,* named Fairchild Semiconductor, would begin to create a whole raft of new silicon-based products, and then shrink and shrink them, and imprint upon them computing abilities that hitherto could be accomplished only by giant machines that occupied entire suites of air-conditioned rooms.

The invention of the planar transistor was by all accounts one of Fairchild's two most important achievements. The man who created the technology that would

---

* The term had not been invented at the time of Fairchild's incorporation (with a five-hundred-dollar investment from each of the eight who left Shockley). Start-ups began to be known as such only in 1970. The founding, in a garage, of Apple Computer in 1976 is a classic example.

then both allow miniaturization to proceed apace and, in turn, lead Gordon Moore to write his famous prediction is now almost entirely forgotten outside the closeted confines of the semiconducting world. Jean Amédée Hoerni, one of the eight who had abandoned Shockley for Fairchild, was the theoretical physicist scion of a Swiss banking family and, at thirty-two, when he joined Fairchild, a devoted rock climber, mountaineer, and thinker.

His elegant invention changed, at a stroke, the way transistors were made. Up until then, they were formed, essentially, mechanically. Tiny grooves were etched into silicon wafers, aluminum conductors were traced into the etched grooves, and the resulting wafers, now with their etched hills and valleys, shaped like western desert mesas (which led these Fairchild products to be known as mesa transistors), were encased in tiny metal canisters, with the three working wires protruding.

These were still somewhat big and ungainly things, and this at a time when, just after the launch of *Sputnik*, the American space industry dearly wanted its new electronics to be tiny, reliable, and cheap. Moreover, the Fairchild mesa transistors were not very reliable: far too often, tiny pieces of resin or solder or dust would be left behind after the etching process and would rattle around in their metal cases and cause the transistors to perform erratically, or not at all. Something was needed that was small and worked perfectly.

The moody, solitary, and austere Jean Hoerni came up with the idea that allowed a coating of silicon oxide on top of a pure silicon crystal to be used as an integral part of the transistor, as an insulator, and with no hills or valleys, no mesas, to give the resulting device unnecessary bulk. His creation, he insisted, would be very much smaller than a mesa transistor, and more reliable. To prove his point, he had a technician create a prototype that was no more than a dot, just a millimeter across, and then dramatically spat on it to show that any kind of human misbehavior would not interfere with its working. It performed flawlessly. It was tiny, it worked, and it seemed well-nigh indestructible—or, at least, immune to insult. It was also cheap, and in consequence, it became Fairchild's signature product from almost that moment on.

It was, however, just one of two Fairchild game-changing products. The other was born of an idea that was doodled on four pages of a company notebook★ by

---

★ Because so many bright people marooned in their offices were doodling ideas in their notebooks, it became customary for the Fairchild company lawyers to demand that the doodled pages be witnessed and signed, to make sure that any ideas that de-

another of the refugees from Shockley, a man named Robert Noyce. His thought was that, now that planar transistors were about to become a reality, might it not be possible to put flattened versions of the *other* components of a full-fledged electrical circuit (resistors, capacitors, oscillators, diodes, and the like) onto the same silicon oxide covering of a silicon wafer? Could not the circuitry, in other words, be *integrated*?

If it could, and if every component were now tiny and, most important, virtually flat, then could the circuits not be *printed* onto the silicon wafer photographically, employing the same principle that was used in photographic enlargers?

The principle of a darkroom enlarger formed the basis of the idea. An enlarger took the negative of a tiny piece of celluloid imagery (say, 35 mm film from a camera) and used its lenses to make a very much bigger version of the image, or an edited part of the image, and print it on light-sensitive paper. This same principle, Noyce

---

served to be patented were credited to the right person. Robert Noyce, for example, witnessed and signed Hoerni's notebook pages relating to the planar transistor. Oddly, though, Noyce's four pages of notes written in January 1959 were never witnessed and signed, meaning that the genesis of the concept of the integrated circuit, the subject of his writing, was never formally agreed upon—anecdotally, yes; but formally and legally, no.

wrote in his notebook, could surely be used backward. A designer could draw onto some transparent medium a large diagram of a piece of integrated circuitry and then, using a device much like an enlarger, but with its lenses refashioned to make its images not bigger but very much smaller, the image could be printed, as it were, onto the silicon oxide of the wafer.

Machines capable of performing such a task, known as photolithography, were already available. Letterpress printers, for example, were employing the idea when, at around this time, they began switching to the use of polymer plates. Instead of using hand-assembled forms of lead characters, a printer could now simply type in a page of work and feed it into a photolithography engine, and out would come a page reproduced as a sheet of flexible polymer. All the letters and other characters, all the *p*'s and *q*'s, would now be standing type high above the polymer plate's surface, ready to be impressed onto paper with a platen press, say, which would give the same look and feel to the resulting page of paper as an old-fashioned piece of handmade letterpress work. Why not modify such a machine to print imagery, of circuitry rather than literature, onto not polymer or paper but silicon wafers?

The mechanics of actually doing such a thing turned out to be formidable—all the imagery was tiny, all the work necessarily demanded the highest precision and the closest tolerances, and the results were minute in aspect and at first imperfect, almost every time. Yet, after months of work in the early 1960s, Robert Noyce and Gordon Moore and Moore's team at Fairchild eventually managed to assemble the congregated devices, and to make them planar—to flatten them and reduce their volume and their power consumption and their heat emission, and to place them together on a flat substrate and market them as integrated circuits.

This was the true breakthrough. Lilienfeld had been first with the idea, in the 1920s; Shockley and his team of Nobel laureates at Bell Labs had taken the first shaky steps; and then, with Hoerni's invention of the planar transistor, with the internals being arranged in thin layers rather than as discrete crystals, suddenly it became possible to miniaturize the circuitry, to make electronics of ever-increasing speed and power and ever-decreasing size.

The transistors in these circuits, with just the application of tiny bursts of power, could be switched on and off and on and off ceaselessly and very swiftly. These new minute baubles of silicon thus became crucial to the making of computers, which make all their analog and later their digital calculations on the basis of a transistor's binary state, on or off—and if the transistors are numerous enough and swift enough in their performances of this task, they can render a computer very powerful, extremely quick, and enticingly cheap. So the making of integrated circuits inexorably led to the making of the personal computer—and to scores upon scores of other devices at the heart of which were ever-smaller and ever-quicker pieces of circuitry, conceived and designed initially by the clever group at Fairchild.

Financially, though, Fairchild performed dismally, not least because other start-ups, such as Texas Instruments,* had the extra cash or a generous parent to allow them to expand into the emerging market. It was their frustration at Fairchild's inability to compete that led the most ambitious of the company's founders to leave yet again, and establish their own firm anew, one that would solely design and manufacture semiconductors. This company, set up by Gordon Moore and Rob-

---

* Texas Instruments also created integrated circuits, but using the bulkier mesa transistors instead of Fairchild's planar versions. Nonetheless, the firm's Jack Kilby won the 2000 Nobel Prize in Physics for his invention. Robert Noyce had died ten years before: Kilby was gracious in his acceptance speech, allowing that Noyce, even though at a competing company, had been the integrated circuit's co-inventor, deserving of the honor, too.

ert Noyce—the "fairchildren," they were called—was set up in July 1968 as Intel Corporation.

Within three years of incorporation, the first-ever commercially available microprocessor (a computer on a chip) was officially announced. It was the Intel 4004, the famous "forty-oh-four." And as an indication of the new kind of precision that was being brought to bear on this new kind of technology, it is worth remembering that buried within the inch-long processor was a tiny die of silicon, twelve millimeters wide, on which was engraved a marvel of integrated circuitry printed with no fewer than 2,300 transistors. In 1947, a transistor was the size of a small child's hand. In 1971, twenty-four years later, the transistors in a microprocessor were just ten microns wide, a tenth of the diameter of a human hair. Hand to hair. Minute had now become minuscule. A profound change was settling on the world.

Initially, Intel's 4004 chip was created privately for a Japanese calculator-making firm named Busicom, which was struggling somewhat financially and needed to lower its production costs, and so thought to introduce computer chips into its calculating engines—and therefore approached Intel. It is part of Intel company lore that at a brainstorming session in a hotel in the old Japanese city of Nara, a woman whose name has since been forgotten designed the basic internal architecture of the calculator in such a way as to positively require Intel, with its unique new miniaturizing abilities, to make the necessary little processing unit.

The calculating machine was eventually created, and launched in November 1971, with advertisements describing it as the world's first desktop machine to use an integrated circuit, a processing chip with the power of one of the legendary ENIAC room-size computers at its heart. A year later, the firm asked Intel to lower its prices for the chips—they were then priced at about twenty-five dollars apiece. Intel said yes, but on condition that it take back the rights to sell its invention on the open market, a stipulation to which the Japanese firm reluctantly agreed. The 4004 was thereafter incorporated into a Bally computer-augmented pinball machine, and was reputedly, but wrongly, said to be aboard NASA's *Pioneer 10* space probe. NASA had thought about using it, but had decided it was too new—and the resulting chipless spacecraft spent thirty-one years after its launch in 1972 wandering through the solar system, its batteries finally giving out in 2003, seven billion miles from home.

The repute of the 4004 spread, and Intel decided that the firm's core business from now on would be to make microprocessors, guided by Gordon Moore's insistence (first published in 1965, six years before his company actually made the first 4004, which hints at a certain prescience) that every year the size of these chips would halve and their speed and power would double. The minuscule would, in other words, become the microscopic, and then the submicroscopic, and then, perhaps, the atomic. Moore revised his prediction after seeing the workings and the challenges of the designing of the 4004, insisting now that the changes would occur every two years, not one. It was a prophecy that has become almost precisely self-fulfilling for all the years since 1971.

And so the near-exponential process of chips becoming ever tinier and ever more precise got under way—with two decided advantages recognized by the accountants of all the companies that decided to make chips, Intel of course included: the smaller the chips became, the cheaper they were to make. They also became more efficient: the smaller the transistor, the less electricity needed to make it work, and the faster it could operate—and so, on that level, its operations were cheaper, too.

No other industry with a fondness for small (the makers of wristwatches being an example) equates tininess with cheapness. A thin watch is likely to be much costlier to make than a fat one, but because of the exponentiality inherent in chip making, because the number of chips that can be crammed onto a single line is automatically squared once you translate the line to a chip, each individual transistor becomes less costly to manufacture. Place a thousand transistors onto a single line of silicon, and then square it, and without significant additional cost, you produce a chip with a million transistors. It is a business plan without any obvious disadvantage.

The measure of a chip is usually expressed by what is confusingly called its process node, which, very crudely put, is the distance between any two of its adjoining transistors, or a measure of the time taken for an electrical impulse to move from one transistor to another. Such a measure is more likely to offer to semiconductor specialists a realistic picture of the power and speed of the circuitry. For an observer outside the industry, it is still the number of transistors on the wafer that offers the somewhat more dramatic illustration, even though a substantial number of those transistors are there to perform functions that have nothing to do with the chip's performance.

Node size has shrunk almost exactly as Gordon Moore predicted it would. In 1971, the transistors on the Intel 4004 were ten microns apart—a space only about the size of a droplet of fog separated each one of the 2,300 transistors on the board. By 1985, the nodes on an Intel 80386 chip had come down to one micron, the diameter of a typical bacterium. By 1985, processors typically had more than a million transistors. And yet still more, more, were to be found on ever-newer generations of chips—and down, down, down the node distances came. Chips with names such as Klamath in 1995, Coppermine in 1999, Wolfdale, Clarkdale, Ivy Bridge, and Broadwell during the first fifteen years of the new millennium—all took part in what seemed to be a never-ending race.

With all these last-named chips, measuring their nodes in microns had become quite valueless—only using nanometers, which were units one thousand times smaller, billionths of a meter, now made sense. When the Broadwell family of chips was created in 2016, node size was down to a previously inconceivably tiny fourteen-billionths of a meter (the size of the smallest of viruses), and each wafer contained no fewer than *seven billion* transistors. The Skylake chips made by Intel at the time of this writing have transistors that are sixty times smaller than the wavelength of light used by human eyes, and so are literally invisible (whereas the transistors in a 4004 could quite easily be seen through a child's microscope).

There are still ever-more-staggering numbers in the

works, ever more transistors and ever-tinier node sizes yet to come—and all still fall within the parameters suggested by Moore in 1965. The industry, half a century old now, is doing its level best, egged on by the beneficial economics of the arrangement, to keep the law firmly in its sights, and to achieve it, or to better it, year after year for the foreseeable future. A confident Intel executive once remarked that the number of transistors on a chip made in 2020 might well exceed the number of neurons in the human brain—with all the incalculable implications such a statistic suggests.

Enormous machines such as the fifteen that started to arrive at Intel's Chandler fab from Amsterdam in 2018 are employed to help secure this goal. The machines' maker, ASML—the firm was originally called Advanced Semiconductor Materials International—was founded in 1984, spun out from Philips, the Dutch company initially famous for its electric razors and lightbulbs. The lighting connection was key, as the machine tools that the company was established to make in those early days of the integrated circuit used intense beams of light to etch traces in photosensitive chemicals on the chips, and then went on to employ lasers and other intense sources as the dimensions of the transistors on the chips became ever more diminished.

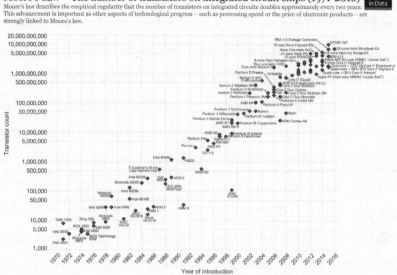

Moore's Law – The number of transistors on integrated circuit chips (1971-2016)
Moore's law describes the empirical regularity that the number of transistors on integrated circuits doubles approximately every two years. This advancement is important as other aspects of technological progress – such as processing speed or the price of electronic products – are strongly linked to Moore's law.

Beginning with the Intel 4004 integrated circuit, which crammed 2,300 transistors onto a sliver of silicon 12 mm wide, and proceeding to today's chips, which contain upwards of 10 billion discrete transistors on an even tinier chip, this graph displays the relentless truth of Moore's law.

It takes three months to complete a microprocessing chip, starting with the growing of a four-hundred-pound, very fragile, cylindrical boule of pure smelted silicon, which fine-wire saws will cut into dinner plate–size wafers, each an exact two-thirds of a millimeter thick. Chemicals and polishing machines will then smooth the upper surface of each wafer to a mirror finish, after which the polished discs are loaded into ASML machines for the long and tedious process toward becoming operational computer chips.

Each wafer will eventually be cut along the lines of a grid that will extract a thousand chip dice from it—and each single die, an exactly cut fragment of the wafer, will eventually hold the billions of transistors that form the nonbeating heart of every computer, cellphone, video game, navigation system, and calculator on modern Earth, and every satellite and space vehicle above and beyond it. What happens to the wafers before the chips are cut out of them demands an almost unimaginable degree of miniaturization. Patterns of newly designed transistor arrays are drawn with immense care onto transparent fused silica masks, and then lasers are fired through these masks and the beams directed through arrays of lenses or bounced off long reaches of mirrors, eventually to imprint a highly shrunken version of the patterns onto an exact spot on the gridded wafer, so that the pattern is reproduced, in tiny exactitude, time and time again.

After the first pass by the laser light, the wafer is removed, is carefully washed and dried, and then is brought back to the machine, whence the process of having another submicroscopic pattern imprinted on it by a laser is repeated, and then again and again, until thirty, forty, as many as sixty infinitesimally thin layers of patterns (each layer and each tiny piece of each layer a complex array of electronic circuitry) are engraved, one on top of the other. When the final etching is done and the wafer emerges, presumably now exhausted from its repeated lasering and etching and washing and drying, it is barely any thicker than when it entered as a virgin wafer three months before, such is the fineness of the work the machine has performed upon it.

Cleanliness is of paramount importance. Imagine what might occur if the tiniest fragment of dust were to settle momentarily on top of the mask where the pattern was to be drawn, at the moment the laser was fired through it. Though the dust particle might well be invisible to the human eye, smaller than a wavelength of visible light, once its shadow passed through all the lenses, by way of all the mirrors, it would become a massive black spot on the wafer, with the result that hundreds of potential chips would have been ruined, thousands of dollars' worth of product lost forever. This is why everything that goes on within the ASML boxes does so in warehouse-size rooms that are thousands of times cleaner than the world beyond.

There are well-known and internationally agreed standards of cleanliness for various manufacturing processes, and while one might suppose that the clean room at the Goddard Space Center in Maryland, where NASA engineers assembled the James Webb Space Telescope, was clean, it was in fact clean only up to a standard known as ISO number 7, which allows there to be 352,000 half-micron-size particles in every cubic meter of air. Rooms within the ASML facility in Holland are very much cleaner than that. They are clean to the far more brutally restrictive demands of ISO number 1, which permits only 10 particles of just one-tenth of a micron per cubic meter, and no particles of any size larger than that. A human being existing in a normal environment swims in a miasma of air and vapor that is five million times less clean. Such are the demands of the modern integrated circuit universe, where precision seems to be reaching into the world of the entirely unreal, and near-incredible.

With the latest photolithographic equipment at hand, we are able to make chips today that contain multitudes: seven billion transistors on one circuit, a hundred million transistors corralled within one square millimeter of chip space. But with numbers like this comes a warning. Limits surely are being reached. The train that left the railhead in 1971 may be about to arrive, after a journey of almost half a century, at the majesty of the terminus. Such a reality seems increasingly probable, not least because as the space between transistors diminishes ever more, it fast approaches the diameter of individual atoms. And with spaces that small, leakage of some properties of one

The main mirror for the James Webb Space Telescope. At more than twenty-four feet in diameter, it will, from its location a million miles from Earth, vastly increase our ability to peer into the very edge of the universe, and at the time the universe was forming. It is due to be launched in 2019.

transistor (whether electric, electronic, atomic, photonic, or quantum-related properties) into the field of another will surely soon be experienced. There will be, in short, a short circuit—maybe a sparkless and unspectacular short circuit, but a misfire nonetheless, with consequences for the efficiency and utility of the chip and of the computer or other device at the heart of which it lies.

Thus is the tocsin being sounded. And yet, to a true chipaholic—or to a true believer that the world will be a better place if Moore's law is rigidly observed and its predictions are followed to the letter—the mantra is a familiar one: "Just one more. One more try." One more doubling of power, one more halving of size. Let *impossible* be a word that in this particular industry goes unspoken, unheard, and unheeded. Molecular reality may be about to try to impose new rules, but these are rules that fly in the face of everything that has passed before, and their observance denies the computer world the role of ambition, and of having its reach extend, as it has for all the years of its existence, well beyond its grasp.

And so the chip-making-machine makers (particularly those in Holland, who have invested billions in this industry, and keenly want and need to preserve their investment) are now doing their level best to comply, to fulfill the wishes of the makers of what some might think are technically impossible dreams. Their new generation of devices does appear to have the ability to let the chip makers go even smaller, even beyond what seems to be possible, or prudent, or both.

The new machines no longer employ visible-light lasers, but what is known as extreme ultraviolet (EUV) radiation, and at a specific wavelength of 13.5 billionths of a meter. This would enable, in theory, the making of transistors down to atomic scale, to edge-of-the-seat, leading-edge, bleeding-edge ultrasubmicroscopic precision, while maintaining some kind of commercial edge, too.

Dealing with EUV radiation is far from easy. It is radiation that travels only in a vacuum. It cannot be focused by lenses, and it won't work with mirrors as mirrors are generally known, but only through costly, many-layered devices known as Bragg reflectors. Moreover, EUV radiation is best produced from a plasma, a high-temperature gaseous form of molten metal that can best be procured by firing a conventional high-powered laser at a suitable metal.

An American company (which ASML subsequently bought) had already developed a unique means of producing this particular and peculiar type of EUV radiation. Some said the company's method verged on the insane, and it is easy to see why.

Extremely pure metallic tin is heated until it becomes molten, and the hot liquid is then squirted out into a vacuum chamber in a tiny jet stream that looks continuous but is in fact composed of fifty thousand droplets moving past each second. The droplets themselves are then hit with light from a first laser, which pancakes each one, making a larger surface area for a second and very powerful carbon dioxide laser to irradiate each flat droplet—each of which turns instantly into a superheated plasma that emits a second jet stream of the wanted extreme ultraviolet radiation. (The bombarded droplets also produce fragments of waste tin, which might solidify were it not for a conveniently sited jet of hydrogen gas that casually brushes them out of the way.)

The EUV radiation that is born in this Hadean environment is then passed through the intricate masks on which the transistor arrays are drawn, that is, the new and ultra-tiny integrated circuit, after which it is moved down a staircase pathway of Bragg reflectors, each made to formidable optical precision, and onto the silicon wafer itself, to begin its work at mechanical tolerances of as little as seven-, maybe even five-billionths of a meter. If everything works properly—and at the time of this writing, it seems to be—then the first of these supercomplex chips, made in this bizarre manner, will be on offer from 2018 onward. And Moore's law, by then fifty-three years old, will prove to have kept itself on target, again.

Yet, all are asking, for how much longer? The use of EUV machines may allow the law's continuance for a short while more, but then the buffers will surely be collided with, at full speed, and all will come to a shuddering halt. The jig, in other words, will soon be up. A Skylake transistor is only about one hundred atoms thick—and although the switching on and off that produces the ones and zeros that are the lifeblood of computing goes on as normal, the fact that such minute components contain so very few atoms makes the storage and usage of these digits increasingly difficult, steadily more elusive. There are plans for getting around the limits, for eking out a few more versions of what might be called "traditional" chips by, among other things, making the chips themselves increasingly three-dimensional—by stacking chip on top of chip and connecting each by forests of ultraprecisely aligned and very tiny wires. This would allow the number of transistors in a chip to keep on increasing for a while without our having to reduce the size of individual transistors.

And there are other materials, other architectures. There is talk of using the curious one-molecule-thick substance graphene, a filmlike, two-dimensional form of pure carbon, for the making of chips. Molybdenum disulfide, black phosphorus, and phosphorus-boron compounds are also being spoken of as alternatives to silicon, as means to keep the juggernaut of ever more miniaturization trundling along, to achieve whatever purpose is demanded of it. The ever-more-alluring field

of quantum computing, which uses the weird ambi-
guities of the subatomic world, as described by Werner
Heisenberg in 1927, as the basis for its abilities, is being
touted as the next step.

Yet, down at this level, measurement becomes increasingly fluid, ambiguity transcends accuracy, precision wanders into the world of paradox, limits become
meaningless, numbers vanish into a quantum-infused
mist—except that there are some real numbers to be
taken seriously. Perhaps most important, there is the
so-called Planck length, the fixed and calculated dimension at which classical ideas of space-time begin to
evaporate and the very idea of physical size becomes
meaningless.

This length has an actual value—or, at least, it has a
value if you believe that the two sure constants in our
known universe, the speed of light and Newton's gravitational constant, are immutably constant themselves.
The Planck length has been worked out as 0.000 000
000 000 000 000 000 000 000 000 000 016 229 (38)
meters, and so is about twenty decimal places smaller
than the diameter of a hydrogen atom. And once you
have that distance, you can work out an extent of
time—if the same constants are similarly immutable,
that is. And so the time it would take for a photon to

journey through a Planck length can be calculated, and has been: the best estimate of this minute expanse of temporal extancy is 5.39×10^{-44} seconds.

And it is at this point that the story of precision becomes, quite literally, topsy-turvy. It becomes wholly impossible to go down, down, down beyond a certain point. Though techniques being studied at some of the national metrology centers and in a few high-octane national and university laboratories around the world allow for some penetration of the atomic limits—a technique known as light squeezing, for example, allows some actual measurement (rather than calculation, which is the basis of those two immensely small numbers just given) of subatomic-level dimension—there is a near-universally acknowledged limit below which things are unmeasurable, and therefore unmakeable.

Going down into the world of the near-atomically minute may have real and proven limitations, but at the other end of the spectrum there are still possibilities. The making of ultraprecisely finished devices and instruments still has validity at the other ends of this topsy-turvydom of extremes. It has value when it comes to the examining of the faraway—as with the James Webb Space Telescope, precisely honed to gaze into the edge of the universe. It has use and validity, too, in the examination of the big cosmological questions that haunt our modern imaginations.

This is why the most exacting limits of precision engineering are now being tested in the construction of the giant instruments at the LIGO sites in Washington State and in Louisiana, and that are about to be built in the plains of western India. LIGO may be the scalar opposite of the integrated circuit, being vast in every sense, the one extending across kilometers and the other covering mere nanometers, but much the same purity and exactitude obtain in the making of both, and perhaps even more dramatically so in the lonely outposts where LIGO is based, and from where it examines one of the enduring and fundamental questions of our cosmos.

Einstein predicted more than a century ago that faraway cosmic events could trigger ripples in the fabric of space-time—gravitational waves, he called them—and that they would change the shape of planet Earth if they ever passed by or through us. The LIGO sites were established to see if such infinitesimal changes in the world's shape were measurable, whether they did in fact exist.

To demonstrate and prove such a tiny change in the

shape of the planet required the building of an enormous and ultrasensitive interferometer. Hence, in 1991, the birth (or, more accurately, the government's funding approval) of LIGO, the Laser Interferometer Gravitational-Wave Observatory, within which are components that lay claim to being the most precise objects ever made by humankind, and that demonstrate just why the utmost precision is needed to examine or create not only at the near-atomic limits of minuteness, but also at the massive scale and near-endless distance of the objects far off in the outer universe.

A classic interferometer uses a powerful light of a pure and known color of which one knows the wavelength. That light is shone through a lens toward a device, basically a half-silvered mirror, that splits the beam in two, exactly. These two tubular beams of pure red light are then directed along paths that are exactly ninety degrees from one another, and toward mirrors that will reflect the beams back toward the first splitting mirror, where they now recombine and are superimposed upon each other as they are directed toward a detector.

If the beams are of exactly the same length, the circular image of the recombined red light will be amplified; the light will be as bright as it was before its beams were split in two. On the other hand, if the two beams differ in length, they will destructively interfere with one another, and the detector will register rings of color that will tell the observers and analysts by how much the difference is.

LIGO is very basically an experiment that employs a pair (soon to be a trio) of enormous interferometers of this quite simple design. Anyone who has used an interferometer would easily recognize, if flying five miles high over the central desert of Washington State, or over the lush forests of south central Louisiana, the two LIGO instruments for what they are: the two long arms at precisely ninety degrees, the building where the two arms meet and where the splitting mirror must be, the extensions and smaller structures where the laser light source is housed and where the detectors and analytical devices are situated, desert scrub up north, beech-magnolia woodlands deep down in Dixie, each suggesting placid and undisturbed nature. The long die-straight pathways cut across their landscapes look Nazca Line–like, stunning in their incongruity.

The purpose of the LIGO experiments is to determine if those two long arms at each observatory change their lengths relative to one another—for, if they do, to the tiniest degree imaginable, then there is a chance that it was the passage through the planet of a gravitational wave that made them do so.

Down at ground level, the instruments are industrial-scale behemoths, with the arms (basically subway-size tubes that stretch into the invisible distance) connected where they join to congregations of humming engine work and electronics of bewildering complexity. Technologies that employ engine oil and technologies that employ silicon perform here in perfect symbiosis. So vacuum pumps pump, laser generators generate laser light, servomotors make microscopic tweaks, and computers in control rooms work through numberless days and nights to interpret the data that stream in as the beams race hither and yon, back and forth, hundreds of times each second, between the mirrors, all in the faintest imaginable hope that, once in a while, the tubes down which the laser beams are pulsing will change length relative to each other.

And this they did, on September 14, 2015, when observers made their first-ever detection of the phenomenon Einstein had predicted almost exactly a century before. The computers in the Livingston control room noticed it: an aberration, an oddity, a variance in the signal, at 05:51 on that Thursday morning, half an hour before local Louisiana sunrise, and with the bayou alligators still asleep. The observers there may have been weary, but this being part of a vast network of participants in what is known as the LIGO Scientific Collaboration, others around the world more bright-eyed and bushy-tailed noticed it, too, at more propitious hours. Back in Hanford, Washington, it would have been 03:51, the dead of night; but in Leibnitz, it was 12:51; in Delhi, 17:21; in Tokyo, 20:51; and at Monash University, in Melbourne, 22:51 in the late evening.

And in every squirrel hole out in the wide beyond, people noticed it. A sudden uptick in a signal was noted in Livingston and was duplicated exactly by the detectors at Hanford. Not that all the detectors were switched on: the observatories were in the middle of an engineering run, when for many months at a time the various components are sedulously checked for precision and accuracy. Normally—not that there is much normal in the world of gravitational waves—observers look out only during observing runs. Yet because nothing had been seen or heard during all the runs of the previous thirteen years—the first basic LIGO was built in the late 1990s and started looking for waves in 2002—and with hundreds of millions of dollars of taxpayer treasure having been spent, with nothing to show for it, there was a sense of, if not quiet desperation, then at least institutionalized eagerness for a result.

So, when the first message came in from a middle-of-the-night observer in Pasadena, headed "Very Interest-

LIGO has two observatories in the United States, one in Louisiana and this one, seen from the air, in the desert of central Washington State. A third is being constructed in an arid region in western India.

ing Event on E[ngineering] R[un] 8," the community pricked up its ears and, as one, shifted into skeptical overdrive.

This could not possibly be, they said. The equipment was in mid-shakedown mode, they said, with machines certain to throw up spurious data from time to time. Besides, part of the system set up to avoid jumping the gun had observers and machines firing off what are called injections, that is, anonymous false results injected, as it were, into the system to keep all the astrophysicists on their intellectual toes.

Days went by, then weeks and months, during which time people around the planet were canvassed. Did you send out an injection? each was asked. And as each responded in a cascade of negatives, and as the results from the two observatories and from other, smaller centers were parsed over and over by analysts and mathematicians of ever-increasing skill and learning and wisdom, the skepticism gradually fell away. The LIGO meisters realized they had a story on their hands. They presented a scientific paper in *Physical Review Letters*, and then, at a crowded press conference in Washington, DC, on February 11, 2016, made an announcement that would shake, or at least stir, the scientific world—and much of the lay world besides.

After a courteous introduction by the director of the National Science Foundation (which took the greatest series of financial risks in its history by committing some $1.1 billion over the forty years since the project commenced), it fell to LIGO's then-director, David Reitze of Caltech, with his astrophysicist colleague Kip Thorne an avuncular presence beside him, to make the formal announcement: that by using *the most precise measuring equipment ever built*, gravitational waves had now been discovered, or more accurately, their presence had been inferred.

"We have done it," he said, and the room erupted in applause. A new era in astronomy had commenced, a new means of discovering the magical complexities of the universe. And a peaceful new era, to boot. It was a moment, someone said, akin to that of Galileo's first looking through his telescope four hundred years before. There were tears of pleasure and pride.

There is an irony immediately apparent to anyone who has been up close and personal with, on the one hand, the 160-ton ASML machines in Holland, which allow for the placing of seven billion transistors onto a wafer of silicon no larger than a fingernail, and the airline hanger–cum–train station vastness of the LIGO machinery that has been established to detect what one author has called "gravity's whispers."

Both sets of machinery have been designed to deal with the tiny, the faint, the microscopic, the atomic, the cosmic—yet both sets of machinery are so Victorian-grand in design and so magisterial in scale, far bigger than the great machines of the past, those that dealt with steam and iron and lathes and screws and governor wheels and flywheels and heat and incessant noise and shuddering vibrations, back when precision was at its vague beginnings. Where precision once employed small machines to construct big things, it now employs big machines to create, or to detect, tiny ones.

There is a further irony.

The first-ever device to call itself precise was a cylinder, bored from a block of solid metal by a Cumberland ironmaster in 1776, specially made for use in James Watt's steam engine, and at the start of the Industrial Revolution. Now, the component at the heart of what LIGO's David Reitze publicly described as "the most precise measuring instrument ever built" is a cylinder, too. Unlike Wilkinson's, this one is solid, a forty-kilogram cylinder known as a test mass and made of fused silica that reflects all but one of every 3.3 million photons that hit it. The silica is tooled and lapped and polished to an immaculate flatness. It is suspended in a cradle by a network of 400-micron-thick silica filaments, and is balanced by an array of weights of glass and metal and magnets and coils that will allow it to be tested and measured by the laser that will hit it 280 times each fraction of a second, in order to measure the distance of the length of the tube at the end of which it lives, and which thereby detects whether a gravitational wave has passed through—as has happened so far four historic times.

John Wilkinson's cylinder fit inside James Watt's steam engine with a degree of precision amounting to the thickness of an English shilling, about one-tenth of an imperial inch. Such precision had never been achieved before, but after that, the world never once looked back.

Two and a half centuries on, and the engineers at LIGO have also made their test mass as a cylinder. This

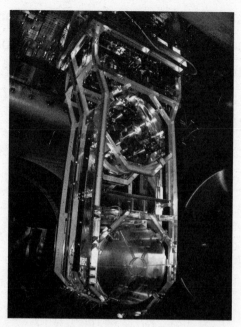

LIGO's precisely crafted fused-silica "test mass" (very basically, an ultraprecise mirror suspended inside a sophisticated damping system) reflects beams of high-intensity laser light that have been shot at it down a 4-km pure-vacuum tunnel in such a manner that it can detect microscopic changes in the tunnel's length, and thus prove the existence of gravitational waves. At the time of this writing, LIGO has proved the existence of four such waves.

one was constructed out of fused silica—a pure form, effectively, of sand, of as elemental a substance, literally and metaphorically, as the iron that was used by John Wilkinson.

The test masses on the LIGO devices in Washington State and Louisiana are so exact in their making that the light reflected by them can be measured to one ten-thousandth of the diameter of a proton. They can also compute with great precision the distance between this planet and our neighbor star Alpha Centauri A, which lies 4.3 light-years away.

The distance in miles of 4.3 light-years is 26 trillion miles, or, in full, 26,000,000,000,000 miles. It is now known with absolute certainty that the cylindrical masses on LIGO can help to measure that vast distance to within *the width of a single human hair.*

And that's precision.

Chapter 10
On the Necessity
for Equipoise

The test of a first-rate intelligence is the ability to hold
two opposing ideas in mind at the same time and still
retain the ability to function.

—F. SCOTT FITZGERALD, *The Crack-Up* (1936)

And yet. The ever-increasing degree of precision
that defines so many of the perfectly ordinary
items that now surround us—and which is supposedly
of such vital importance to the pursuers of today's scientific truths—prompts a cascade of philosophical questions. Is such a wish for perfection truly an essential to
modern health and happiness, a necessary component
of our very being? Do the benefits it provides clearly

outweigh the shortcomings that so clearly accompany its stealthy recent insertion into human life and society? Are we a happier and more contented collective of souls for possessing it and employing it in our everyday? Should we worship and revere and give thanks to all those of the past—Wilkinson, Bramah, Maudslay, Shockley, and their like—for blessing us with their notion of the need for endlessly improving exactitude?

And still further, might there be a gathering of people—a society or a country somewhere in the world—who hold to a subtly different perspective on precision's advantages, who question it as an ideal for aspiration? Might there be a people given to a deep-seated appreciation of the polar opposite of precision—a people who display a real affection, in other words, for the imprecise as well? A people who can perhaps hold dear the two ideas simultaneously, and yet retain, at all levels of society, a keen ability to function?

Japan, I would argue, is just such a place.

This is a country known for its rigorous appreciation of the perfect, both today and in antiquity. The old temples of Kyoto, perhaps most famously, present an immense treasure-house of the architecturally impeccable, with every last beam and finial and spire and wooden gate designed and carved by ancients for whom perfectionism was deliberately meant as an enduring essential, and whose legacy still prompts silence and awe among those fortunate enough to witness it.

And as with the ancient, so too with the modern. Most now think of modern Japan as dominating today's world with her expertise in the making of objects of unyielding accuracy—of lenses immaculately ground and polished, of cameras fashioned to tolerances unattainable by most other manufacturers, of engines and measuring devices and space rockets and mechanical watches of a quality envied by all others—the Germans and the Swiss most notably—and for whom precision is a byword. In Japan it is more: precision in all things—not least in everyday railway services of such legendary punctuality that an apology had to be offered late in 2017 when an express left twenty seconds *early*—can be thought of as part of the national religion.

And as Kyoto amply displays, there is nothing new about this reverence: the blade of a centuries-old samurai sword is to many Japanese no less sublime a piece of engineering than are the more modern products of firms like Nikon and Canon and Seiko and Mitutoyo and Kyocera. Might it be possible that in Japan the very highest esteem was accorded both to the machined accuracy of today and to the hand-created craftwork of old?

So I went off East, to investigate the possibility. Once

in Tokyo I lit out for a pair of towns in the north of the country, to explore the conundrum. I put up at a hotel close to Tokyo Station to undertake two train journeys out into the Honshu countryside. The first was to visit the Seiko wristwatch factory in the sprawling city of Morioka, and where I suspected I might get some kind of answer.

At the main railway station in Morioka, a northern Japanese city of a quarter of a million nestled under the slopes of a classically shaped volcano named Mount Iwate, there are gift shops where you can buy examples of the most revered product of the region, a bulbous, black hammered-iron teakettle known as a *tetsubin*. Local ironworkers have been casting and pummeling out these utensils for centuries, a reminder that, in Japan at least, beauty is manifest in the mundane, the everyday.

For although so much of today's Japan is steeped in the technological wonders of high-precision modernity—the gleaming high-speed bullet trains are a familiar example, being impeccably made and smoothly run, their workings silent, reliable, safe, and fast and their schedules invariably kept—a sizable fraction of the Japanese people remains vocally and demonstratively proud of its homage to craft, fervent in its admiration for those who make, sell, buy, collect, or simply choose to own objects of great and classical beauty, no matter how outwardly ordinary and no matter how imperfect. The quality and design of the handmade *tetsubin* of Morioka are known the length of the land, and all who see a new-bought example will cluck approving sounds, and well know where it is you have been visiting.

Yet that is a holdover from yesterday. In more modern times, the city of Morioka has become known for another, more up-to-date product, one that reflects, in the same way as do the hand-hewn teakettle and the precisely wrought railway train, the curious duality that Japan displays in its esteem for manufactured items of rare quality. Morioka has since the last war been the manufacturing headquarters for the Seiko watch company—and the duality of both effort and attitude toward this firm's products can be seen plainly, and in vivid demonstration, on the adjacent sides of a single unadorned wall on the second floor of the main Seiko factory building.

The origin myth of the company is a beguiling one. Its founder, Kintaro Hattori, was born in central Tokyo in the late nineteenth century and grew up in a nation that was undergoing a swift and profound change—and was himself, in consequence, influenced by two very different sets of manners and customs. When he was born, in 1860, the emperor Mutsuhito★ was still a shadowy figure cloistered scores of miles away in Kyoto, and the shogunate still ruled from the Japanese capital, which at the time was named Edo. By the time the boy was eight, however, all Japan had changed, was now stumbling toward modernity: the last shogun had abdicated; the emperor had moved to what was now to be called Tokyo, the eastern capital; and reform and modernization (which, in many senses, meant at least temporary Westernization) were on all sides.

Among these reforms was a subject that the teenage Hattori found especially fascinating: the passage of time. The boy had developed an infatuation with clocks, a topic which in the Japan of the day was a matter of exceptional complexity. For Japanese timekeeping was unique. Clockmakers had learned the basics of mechanical horology from visiting Jesuits, but these clerics threw their hands up in puzzlement at the fugitive nature of the local version of timekeeping. Hours in

---

★ An emperor's surname ceases to be used after his death, and posthumously, the name of his reign era is applied instead: so Mutsuhito becomes Meiji; Yoshihito became Taisho; Hirohito is now referred to as the Showa emperor; the current emperor, Akihito, will become the Heisei emperor when he dies or abdicates.

Old Japan were of varying lengths. Clocks' bell strikes were, by Western standards, strangely disarranged: six peals at sunset, nine at midnight, eight and then seven in the moments before dawn. There were different periods of time, too, that depended on the seasons and that required at least two balance mechanisms within each clock, and several faces. Even more faces (as many as six) were needed once Western time began to infiltrate the systems of old, in the days when reformists wanted to be able to tell their hour at the same moment as old-timers required to know theirs. The young Hattori, an apprentice to a Ginza clockmaker from 1873 on, was thus pitched into a ferment of variant accuracies and clashing systems that would serve him in later life far better than, at the time, he could have imagined.

For, by 1881, when, with savings and a little family money, he put down a deposit on the rent for a small watch, clock, and jewelry shop in Kyobashi, not far from the brand-new Tokyo Station—railways had come to Japan in 1872, with a British-built line from Tokyo to Yokohama, nine trains daily—Japan was starting to embrace Western timekeeping standards, too. Almost from the day he opened his store, Hattori would happily accept the older *wadokei* clocks that customers brought in for repair, but he would much more happily sell the clocks and pocket watches that displayed the twelve-hour units and sixty-second minutes of the West, and that, to the young man's good fortune, suddenly became all the rage. There might not yet be much money about, but among the Tokyo middle class, there were usually sufficient funds for a pocket watch, and most men of affairs, who were starting to wear Western clothes now, in contented defiance of the shoguns' old customs, liked to be able to pull a fob watch from their waistcoat pocket and tell the time Western-style.

K. Hattori and Company prospered. Within four years, Hattori was importing the most sophisticated of Swiss and German clocks and pocket watches. He set up a company to make timepieces himself, and called it Seikosha, or (by some translations) "the House of Exquisite Workmanship." With careful investment and slow and steady expansion (and a commitment to the business philosophy of vertical integration, whereby a firm owns or has control of most of the companies that provide its parts or raw materials), it can fairly be said that Hattori flourished.

To recount his rise leaves one almost breathless. He set up an American-style factory for the mass production of clocks, employing the same principle of interchangeable parts that had been born in New England two centuries before. By 1909, Hattori's concept of vertical integration was refined to the point where every single component of every single timepiece was made by a firm he owned, just as they still are today. By the turn of the century, his company had become the biggest mass producer of clocks in the country, and was by then exporting, mainly sending Japanese-made wall clocks to China. After clocks came the production-line pocket watches, most famously in 1910, with a line that some might now regard as ominously prescient, it being called the Empire. Then, in 1913, came the more innocent-sounding Laurel, the company's first small and rugged watch, designed to be worn on the wrist, an advantage for soldiers, allowing them to time their simultaneous battlefield risings from the trenches.

Hattori's showcase for his products was a huge main store and showroom that he had built in Ginza, a shopping district in Tokyo, and that sported, perhaps for the first time in Japan, a clock tower, Hattori believing in the PR advantage to be gained when passersby glanced up and saw the name "Hattori and Company" each time they checked the hour.

Yet, like so much of Tokyo, this grand structure was comprehensively ruined in the Great Kanto Earthquake (and subsequent fires) of 1923, whereupon Hattori decided not just to rebuild but, according to today's Seiko management, to replace every single one of the 1,500 pocket watches he then had in his repair shop. A

Seiko—the name in Japanese means "exquisite workmanship" or by some translations "precise"—invented the quartz watch in the 1960s. The clock over one of the firm's early twentieth-century buildings, now a landmark department store, in Ginza, a district in central Tokyo, is said to be connected to an atomic clock and offers the exact time to millions of commuters and shoppers passing beneath it.

mass of coagulated metal in a display case in the company museum in northeastern Tokyo is purported to be the melted remains of watches then under service; all, it is said, were replaced free of charge. The rebuilt Seiko headquarters still stands today, on what has become one of the very busiest of corner blocks in Ginza, and though it was long since sold to a department store, the tower, very much a local landmark, with its boldly illuminated clock, is under perpetual contract to display the name "Seiko." That name had replaced the name "Hattori and Company," which in turn had replaced "Seikosha," a name that enjoyed only a brief run. "Seiko," ever since, has been deemed sufficient.

And because Japanese Railways had shortly before selected Seiko pocket watches as the official timekeeper of the country's vast and enviably punctual transport network, and because all watches in Ginza and beyond are still checked against the clock that looms over the elegant Wako department store—Gucci to the left, Mikimoto pearls to the right—one can fairly say that all Japan now runs to Seiko time. Not for nothing is the firm's name taken by many Japanese to mean "precision," for there surely cannot be a more precise country, anywhere.

The duality, however, remains. It is a conflict that seemingly lies somewhere deep down and unspoken, buried in the Japanese psyche, between, on the one hand, a perceived modern need for the perfect and, on the other, a lingering fondness for the imperfect, and with amiable disputes about the weight that society accords each. There is a Japanese term for the liking for the natural and the ragged and the undermachined:

wabi-sabi, an aesthetic sensibility wherein asymmetry and roughness and impermanence are accorded every bit as much weight as are the exact, the immaculate, and the precise. And that is precisely what I had come north to explore: that other perspective, whether precision itself is a force for universal good, or whether there is, in essence, a third way.

It was within Seiko particularly that this genial argument became fully exposed, and it did so with one of the firm's—one might say with one of the twentieth-century world's—great inventions. For it was the quartz digital watch, launched by Seiko as the Astron on Christmas Day 1969, that brought this divide fully into the open.

Quartz is a crystal that will oscillate dramatically when placed in an enveloping electric field, and moreover, it will do so by an exactly known number of vibrations a second. It can thus be easily adapted to display with great accuracy the passage of time, and has been used in timekeeping since the discovery of the phenomenon in the late 1920s—although, in the years immediately after the concept's discovery, such clocks as were made had to be contained in boxes that were at least the size of a telephone kiosk.

Seiko was, however, secretly experimenting with

miniaturizing the technology in the 1950s, under the unimaginative code name 59A. In 1958, the firm managed to supply a quartz clock for a Nagoya radio station; it had to be housed in a case the size of a filing cabinet. By the early 1960s, however, Seiko quartz clocks were small enough to be installed in the cockpits of the first generation of bullet trains. By 1964, when Seiko won the timekeeping contract for that year's Olympic Games, there was growing confidence that, sooner or later, the engineers would produce a movement small enough to fit on a wristband—as indeed they did, just five years later. The Astron, with a pleasingly retro face and by then with a gearless, springless, and almost wheelless digital interior, was everything that was expected: an inexpensive, unbreakable, unshockable, heat-resistant, waterproof, and uncannily accurate wristwatch that, for a while, was the most precise timepiece ever made.

It was unshockable only in a purely mechanical sense, for its introduction sent economic and social shock waves coursing through the world's watchmaking community. Within no more than five years, it very nearly brought the Swiss industry to its knees. Suddenly, no one seemed to want to buy a heavy, noisily ticking device that had to be adjusted every day to keep it to time. Instead, and for much less of an outlay, you could buy a watch that never had to be wound and that had, instead of a dial with hands, a display of an ever-unrolling cascade of numbers that told the time to fractions of a second and with accuracies that had hitherto been known only in laboratories. Before what became known in horological circles as the 1969 quartz revolution, or shock, or crisis, there had been sixteen hundred Swiss watch houses. By the end of the next decade, there were only six hundred, and the workforce had been cut to a quarter of its former level.

Seiko was, however, never able to win a patent for its invention, and its own scientists would readily admit that the quartz timekeeping movement was a child of many parents. The firm was quite content to allow the stricken rest of the watchmaking world to play catch-up—as indeed it did. The arrival of the Swatch phenomenon in 1983 brought Switzerland roaring back to life, but by this time, Seiko was well established, pumping out quartz watches at a formidable rate, and making formidable profits in the process.

All this caused the firm's management—the new generations of the Hattori family were still involved, though in a respected supervisory role, as Seiko had gone public in the 1980s—to suffer something of a crisis of conscience, a crisis that was brought about by the firm's philosophical reverence for the watchmaker's craft.

And herein lies the dilemma, one that is well illustrated within this one watch company and, at the same time, also reflected and refracted throughout Japan more generally, and which also helps to address the more philosophical problem that I started to think about on my long desert journey from the LIGO site to Seattle Airport.

For might there be in the wider world, in truth, simply *too much precision*? Might today's singular devotion to mechanical exactitude be clouding a valued but very different component of the human condition, one that, as a result, is being allowed to vanish?

On the day I visited the Seiko watch company's principal factory in Morioka in early autumn, rain was falling, and low clouds obscured the usually quite dramatic view of Mount Iwate. A senior executive had accompanied me on the train up from the south, apologizing for weather that I told him I found quite agreeable after the steam bath of Tokyo. The plant is a little way west of

More than twenty-five thousand quartz watches, known for their accuracy and reasonable price, come off a robotic assembly line at the main Seiko plant in Morioka, northern Japan, every day.

town, in a bamboo park, and the trees dripped softly in the cool drizzle, small pathways vanishing alluringly into the mist.

The factory is modern, stark, unadorned, peaceful. Down at reception level, and in the various rooms where I was taken for briefings, all was unusually quiet, almost as though the plant were on holiday and people had been brought in simply to speak to me.

I needn't have worried. One floor above, where the watches were made, there were people and machines in abundance—though a still very serene abundance, with not a room in which earplugs or masks were ever needed, and everywhere an overarching impression of silence, cleanliness, and efficiency. It was more like an academy than an industrial plant, more a cathedral to the watchmaking religion than something as vulgar as a factory.

A quartet of escorts took me first to the electronic side of the plant, to see where the quartz watches were fashioned. There is a long corridor with glass picture windows through which visitors can observe the long production line, all at waist height, where the components are assembled by robots. The line itself snakes around the warehouse-size room, different portions of the room assigned to different models of watches, but the processes of making them all essentially the same.

Components are fed in from hoppers onto tracks and are then, like train cars being inserted onto a moving railroad, injected into the passing line at just the moment they are required, such that once weight sensors detect their presence on the blank, the tools at this or that part of the line perform the tiny tasks that secure each particular component to its precise position in the watch. The blank with its first part is then moved to a station where a second component is added, and then to a third, and so on and on. The long snake of machinery is monitored by young men and women in white gowns—the room is kept as free of dust as possible—who make an adjustment here or add a droplet of lubricant there, bending down slightly every few moments to minister to the engine work of the never-stopping line.

The line runs ceaselessly, day and night. Rather more than a thousand watches are made every hour, meeting the endless demands of the enormous export market Seiko has established, and from which it now makes the greatest portion of its profits. The vision of all these machines, like an enormous model railway layout, humming and whirring, clicking and whooshing and squealing as they cut, pressed, heated, scored, drilled, removed burrs, tightened screws, fastened faces to mechanisms, inserted glass onto dials and straps into brackets and completed watches into boxes, was mesmerizing indeed—though, in truth, it seemed to have little to do with actual watchmaking, and I suspect my hosts could discern a vague ennui in the expression on my face. "Behind this next wall," an escort said with a smile, "there will be what you want."

In 1960, when the company was still making mechanical watches only, it created a top-of-the-line model it called the Grand Seiko. The watch was handmade to exacting standards, and it was old-fashioned, with a design that was manifestly not artificially retro but that looked the way it did simply because its designers were old-fashioned men, too. It sold well, it won all manner of certificates from a somewhat condescending Swiss judging body, but it was never released overseas and remained almost wholly unknown beyond Japan.

Then came the revolution. Seiko invented the quartz watch in 1969, and put the Astron and its successors into full-scale production, and found that, as a result of its immediate success, it had been somewhat hoist with its own petard. The Grand Seiko mechanical watch became an immediate dud. Price was one factor. Accuracy was another—a quartz watch kept time to mere seconds a year, while a mechanical movement, much costlier, would be lucky not to gain or lose five seconds in a single day. All Japan, and all Seiko, swiftly lost interest in the model, sales plummeted, production was cut, elderly men and women who had been hand-making these watches for years were dismissed, and finally, in 1978, the line was abandoned.

Except that—and this appears to be the decisive moment when a quintessentially Japanese devotion to craftsmanship was allowed to resurface—within a decade, the decision came down from the board to *restart*

On the same floor as the machine assembly of cheap watches, a small team of skilled workers assembles mechanical Grand Seiko–model watches by hand. The team (seen here during one of the compulsory exercise breaks) makes about a hundred watches a day, using components that are all, from hands to hairsprings, made in Japan, also by Seiko.

production. A halfhearted 1980s lipstick-on-a-pig attempt to make a quartz version of the Grand Seiko
fizzled, whereupon the Hattoris realized, and did so
without the dubious benefit of surveys and focus groups,
that Japanese people had a lingering love affair with
handmade mechanical watches, and would pay good
money to support the kind of craftsmanship that would
be necessary to make them again.

Managers in the mid-1980s had retained the names
and addresses of all the watchmakers they had sacked,
just in case they were needed to repair any of the Grand
Seikos then around. The word went out for them to return to work, and they promptly did, in droves. Those
young enough stayed on for such time as they could, and
while working assembling watches by hand once again,
they also trained a new cadre of youngsters—who remain in the plant today, in workshops on the other side
of the factory's second-floor wall.

Not a production line is in sight, nor a robot in view.
Instead, viewable from a large sofa placed in front of
this corridor's picture window are two dozen enclosed
workstations, small ebony-walled carrels each with a
270-degree workbench equipped with every imaginable essential of a modern watchmaker's trade: powerful lights, magnifying lenses, computer screens, racks
of personal tools, tweezers, minute screwdrivers, pin

vises, burnishers, dust brushes, pincers, microscopes, ultrasonic cleaners, boxes of tiny jewels, spindles, gear-wheels, mainsprings, timing devices. All these treasures are arranged with perfect tidiness and ease of access for the man or woman who, in white cotton cap and white cotton gown, sits in his or her custom-made chair, which rises to exactly the correct height for the forearms to be rested and the hands to be as comfortably placed as possible, and makes watches by hand.

When I arrived at the window, every one of the watchmakers was silently peering through the illuminated lens before him or her at some unimaginably tiny piece of a watch-to-be—fully trained watchmakers here work to tolerances of a hundredth of a millimeter, better in some instances. All the pieces, oscillating balances to hairsprings, train plates to escape wheels, winding crowns to pallet forks, are made by hand behind yet another wall of the same building. With tiny tweezers, each of these craftsmen and -women could be seen fitting this piece into this minute hole, into that microscopic space, that tiny threaded notch. Most were bent down, concentrating fiercely on the task at hand. Once in a while, a watchmaker might look up, might glimpse a passing visitor, grin for a second, and then bow down once more and work on.

Every hour, the entire studio breaks for ten minutes

of exercise, and the men and women stand and stretch and limber up for another session hand-making some of the most unassumingly magnificent watches ever made. These watches may not be as famous as Patek or Rolex or Omega, but they consistently win all the Swiss timekeeping awards, and to those who know, they are of peerless quality.

One of the watchmakers came out for a break: he was a slightly chubby, entirely affable man of forty-five named Tsutomi Ito, and he described himself as an expert on hairsprings. He loved the way they rippled sinuously when touched—if, of course, they had been perfectly made. He had been working at Seiko for most of his life, and imagined he would do so until his hands or his eyes gave up under the strain, which they currently showed no sign of doing. He was classified as a meister, one of just two at the plant.

He had begun his career in the electronic watch section, helping to maintain the production line. His ambition had always been to make it to the mechanical watch studio, since human perfection was the essential component here, not the robotic efficiency of the quartz production line. He now completed just two, sometimes three watches a day. In the evenings, he went fly-fishing, and yes, he designed, made, and tied his own flies. He also collected fine wristwatches from around the world. He noticed my Rolex Explorer but would not comment on its quality. Did he like quartz watches? Well, he said, they are a lot more precise than the watches he made. Would he wear one? He shook his head, positively shuddering at the thought. And then he smiled, looked at his own watch, a Grand Seiko diver's watch, mechanical, and stood, allowing how he had now to go back to his workstation. There was a hairspring to adjust, one that was being especially trying. He would like to finish it by closing time, or he would be late getting home. He looked at my Rolex as we shook hands, and gave what I could only suppose was a slightly sardonic smile.

Seiko makes twenty-five thousand quartz watches each day, seven days a week. On a good day, Mr. Ito and his two dozen colleagues who make mechanical watches by hand from Monday to Friday turn out around one hundred twenty. In the reception area, a small glass case showed the very latest models, and a sign noted that, by application to the receptionist, the case could be unlocked, and it was possible to use a Visa card. For a second—for a single tick of a Grand Seiko mechanical—I hesitated. Would you take my Rolex in exchange, I asked, and the team of escorts in the lobby exploded in relieved laughter. I took that as a no. I stepped out into the warm rain and gazed down into one of the bamboo trails, at a view of subtle loveliness that faded into the cool autumn mist.

It was an altogether much less attractive view that greeted me a few days later, when I made a second northward journey from Tokyo, to the coastal fishing port of Minamisanriku. One of the towns that had been wrecked by the Great Tohoku Earthquake and Tsunami of March 11, 2011, and from which, more than six years later, it was still recovering.

Before the tsunami roared in that chill afternoon, Minamisanriku was a prosperous and well-oiled fishing port, if declining slowly in population and importance. Though it stood at the head of a large sheltered bay, few of its fishermen troubled to venture out into the Pacific Ocean itself. There was no need. Just beyond the headland cliffs, the commingling of two marine currents, one warm, the other cold, created a marine environment that was amply suited to a wide variety of harvestable sea creatures.

The local fishermen farmed oysters and scallops, octopus and salmon, and a peculiarly ugly creature called a *hoya*, or "sea pineapple," which has something of a following among the more adventurous Tokyo chefs.

The bounty would be put onto each evening train to the junction at Sendai, and then onto one of the southbound expresses to the city, two hundred miles away: bidders at the Tsukiji morning market would buy it for good prices. Minamisanriku was in consequence well off, contented, and settled—though eternally aware of the ocean beyond the cliffs and the violence it could do. Considerable damage had already been done by a tsunami in 1960. As it had been caused by an earthquake in Chile, the Japanese chose an Easter Island *moai* as an additional town mascot, to act as co-talisman with the more venerable figure of the town octopus.

In no more than one hour on the March Friday in 2011, everything that had for so long been so settled about Minamisanriku was rendered into splintered driftwood, twisted iron, and broken and drowned bodies. Though, outwardly, similar violence wrecked a score of communities up and down the Tohoku coast, that in Minamisanriku had its own peculiar, local poignancy: one tragedy stood out, and made this community's misery more public than many others'. A twenty-four-year-old woman named Miki Endo had been employed to warn the community of the inrushing waters, and on that cold March day, she remained dutifully at her post in the town's Crisis Management

Center as the freezing floodwater rose around her. Just as with the musicians on the *Titanic*, she carried on, sounding the sirens and playing her warning music and broadcasting details of the incoming waves' heights and locations over the municipal loudspeakers, until the water shorted the power supply and the speakers went dead.

Film clips show the waters climbing higher and higher up the center's three stories, until figures can be seen gathering out on its flat roof. A few of them clamber up the radio antennas until only one or two remain; men can be seen holding on grimly and for hours, until the waters begin to drop. Behind them, in one scene, immense gray waterfalls gush through the upper windows of the town hospital, as apocalyptic a vision as it is possible to imagine. But there is silence from the loudspeakers, a lack of sound that tells of the fate of the drowned Endo, who remains the town's local heroine today, for shouting out the warnings until she fell.

The rust-red iron frame of the building in which she was entombed still stands. There is currently a vigorous debate about whether it should stay, as a reminder, like the dome at Hiroshima. Many locals want it torn down. The town has yet to decide.

Endo was but one of some twelve hundred who died

at Minamisanriku, out of a total population of seventeen thousand. The steep hills surrounding the fishing port provided sanctuary for many thousands who either lived there among the pine and cedar and, most crucially, bamboo forests, or else who drove frantically up roads that normally require tire chains in the icy weather—and it did snow that afternoon, though mercifully, only a little. From up high, they watched helplessly as their community was inundated by seven great wave fronts and was methodically wrecked beyond recognition. Then they all came downhill and, by all accounts, patiently and uncomplainingly cleared up the mess and got back to work.

What, one might ask, did they have to work with once they came down from the hills? What remained standing after the waves had stilled? Precious little, for certain, that had ever been made with precision.

Precious little was left in Minamisanriku that had been made of titanium, or of steel, or of glass. Ships with super-precise engine work had been wrecked; cars loaded with exact pieces of instruments and apparatus had been tossed like chaff; electronic devices with microprocessors at their hearts and which held millions of tiny transistors all failed; and buildings such as Miki Endo's were torn apart and twisted and left to rust. The evidence of the impermanence of the precise was everywhere.

The more perfect of the trees, the cedars and the pines—they also were ruined, splintering and collapsing. Many were the human victims who were crushed by their falling trunks, or else were carried along with a floating mass of wrecked driftwood and taken out to sea on the ebb, to be lost forever.

The imprecise, though, was still there. In the forests around town there still were groves of bamboo, growing in abundance. The cedars had gone, splintered to shreds. The pines were devastated. But the bamboo was still there—imprecise, imperfect, but surviving.

Bamboo, used in so many aspects of Chinese and Japanese daily life (as baskets, clothes, tools, fans, shelters, arrows, hats, armor, building material), is a grass, though it appears most commonly as a strong and fast-growing tree. It is renowned for its resilience and flexibility, and it is always certain to grow back and to flourish and then to be useful to mankind for myriad purposes—no matter how many more tsunamis may be inflicted upon it. It bends, it springs back, and it grows again. And in Minamisanriku, it was either still there, bent and bloodied but unbowed, or else it so swiftly reemerged from seed, growing three feet a day once the sun was up and some warmth began to seep back into the springtime earth, that it immediately became of practical use. It is a plant at once mathematically imperfect and yet quite perfectly useful.

As I was leaving New York for Japan in the autumn of 2017, an exhibition was opening at the Metropolitan Museum of Art devoted entirely to the art of bamboo. Most of what was on show (to thousands, for it was a dramatically curated and in consequence a very popular exhibition) was more decorative than strictly practical: many flower baskets and tea ceremony utensils, gift boxes and small items of headgear. But the exhibition also reminded visitors of the existence of what are known as Living National Treasures, Japan's near-unique way of rewarding and recognizing those in society who are creators of the very best of handmade craft.

The very existence of such officially honored artists serves as a reminder that there truly is in this regard an

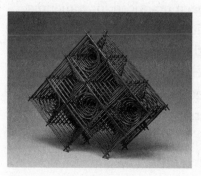

The intricacies of handcrafted bamboo—here a modern Japanese decorative item from an exhibition held in New York City in 2017—demonstrate Japan's pride in the handmade and the imprecise, despite the country's known aptitude for high-precision manufacturing.

ineffable difference about Japan, a singular quality that uniquely marks out the popular attitude toward, in this case, dimensional integrity. For while there is indeed a national reverence in that country for the precise, there is also a formal recognition of the inestimable value to society of craftsmanship, of the true worth of the handmade and the flexibly imprecise.

Living National Treasures represent a corps d'élite of men and women, usually of considerable age, who have over their lifetimes developed and honed skills in such defiantly imprecise arts as lacquerware and ceramics and wood- and metalwork, and who are officially accorded honored status in society.

The virtue central to each of their skills has to be that of patience, both the patience demanded in the learning

Urushi, or handmade lacquerware, is the product of an ancient and respected craft in Japan, one that has been practiced for millennia. The items are made over many months, from the resin of the fiercely protected lacquer tree. So eager is Japan to keep alive the tenets of craftsmanship that it awards to the most honored makers of these beautiful items the title of "Living National Treasure."

of the craft and the patience required in the making of the art.

Urushi, the Japanese name for the ancient craft of lacquerware, for example, offers a perfect illustration of the creation of the imperfect, by way of a skill that has been honed over seven millennia of Japanese history.

The natural material central to the lacquerware art is the highly toxic sap of a tall deciduous tree, *Toxicodendron verniciflluum*,★ which is known principally in China and India but which for centuries has been cultivated in fiercely protected forests in Japan and Korea. Using small blades and buckets and a great deal of care, sap collectors incise tiny feather-like grooves in each tree and collect the drips of sap before each wound heals; the tree is then left unscathed for the following season. Half a cup of sap per tree is the general rule, and each resulting container of the sticky moistness, with pigment adding various shades from deep red to deep yellow to tobacco brown, is tightly sealed until the

---

★ As its new specific genus name suggests—its genus was formerly known as *Rhus*—the tree's leaves are poisonous, afflicting incautious forest workers with dangerous rashes, much like but much worse than those resulting from contact with America's poison ivy, a close botanic relative.

urushi artist calls it up to begin its application, burnishing and decorating.

Generally, wood is used as a base—camphor and cypress wood commonly, air-dried for as long as seven years to ensure no warping or cracking, and then cut and shaped and shaved until it is so thin as to be almost transparent: one can certainly see light and dark through it, or discern the fingers of the artist's hand, if not perhaps read the fine print of the day's *Asahi Shimbun* through it.

Then the lacquer itself is applied to this fragile wooden substrate, painted on with a combination of animal-hair brushes and slender, flat spatulas, done in the thinnest possible of layers, with each stratum left to dry in warm, damp air, both to encourage oxidation and to stimulate the release of the various enzymes that help harden and render permanent the layers, one by one. Maybe as many as twenty layers will be painted on, one atop the other, and smoothed and polished each time, so that each layer is painted onto an unruffled surface, the smoothness of one reflected up onto the smoothness of the next, until a hard, creamy silkiness of texture and surface disguises and also augments the near-invisible structure of the wood below.

More drying, more maturing; polishing with fragments of charcoal and soapstone, chamois leather and clay-soaked silk—the surface now gleams and reflects, though with nothing resembling either glitter or gaudiness, but rather, a near-living texture of a gentle softness, ready only now for the application of the finest paints, or of gold dust, or silver lines, to be finished. It goes without saying that this last decorative process can take weeks or months, as the *urushi* artist makes his ink jar or *bento* box or kettle or tea bowl (tea bowl most especially) into a thing of perpetual elegance, due to represent his country's artistic tradition for centuries to come.

Patience and fine material, amalgamated with the enduring vision of the artist, who now slips into the background and quite deliberately pushes his art to the fore, are the essential elements in the making of the finest of Japanese craftsmanship. And to most cultured Japanese, it matters less whether that art is expressed by way of lacquer or porcelain, through intricately worked metal or delicately carved and jointed and polished wood, than whether it is performed with patience and care and tenderness and—dare one say it?—with reverence and love. The human participation—in no way a dominating participation, for in Japan, the artist seeks only to work in cooperation with his material, and to do so over accumulated amounts of time—is key. No machine will be employed, only well-worn hand tools that have been maintained and perfected for generations. The results define a nation and a people: to see a lacquer tea bowl is to see Japan in all her centuries of dedication to her craft.

And all this craft, in a way, celebrates impermanence. Few other countries in the world make it so abundantly and officially clear that equal weight, respect, and admiration must be accorded both to the precise and to its opposite, to machine and to craft. That respect be accorded to titanium on the one hand and, on the other, to—yes, to such creations of the human mind and hand, to be sure—that most classically Japanese plant, found these days on the hillside of the recovering Minamisanriku and, at the time of this writing, on view as previously mentioned at the Metropolitan Museum of Art: bamboo.

Humankind more generally, obsessed and impressed today with the perceived worth of the finely finished edge and the perfectly spherical bearing and by degrees of flatness that are not known outside the world of the engineer, would perhaps do well similarly to learn to accept the equal significance, the equal weight, of the natural order. If not, then nature will in time overrun, and the green strands of jungle grass—and yes, the green strands of young bamboo—will eventually en-fold and enwrap all the inventions that we make, no matter whether their tolerance is that of the thickness of an English shilling or a fraction of the diameter of a proton.

Before the imprecision of the natural world, all will falter, none shall survive—no matter how precise.

Afterword:
The Measure of All Things

Perfection is the child of time.

—BISHOP JOSEPH HALL, *Works* (1625)

Humankind has for most of its civilized existence been in the habit of measuring things. How far from this river to that hill? How tall is this man, that tree? How much milk shall I barter? What weight is that cow? How much length of cloth is required? How long has elapsed since the sun rose this morning? And what is the time right now? All life depends to some extent on measurement, and in the very earliest days of social organization a clear indication of advancement and sophistication was the degree to which systems of measurement had been established, codified, agreed to, and employed.

The naming of units of measurement was of course one of the first orders of business in early civilization—the cubits of the Babylonians were probably the first units of length; there were the *unciae* of the Romans, the grain, the carat, the *toise*, the catty—and the yard and the half yard, the span, the finger, and the nail of early England.

The later development of precision, however, demanded not so much a range of exotically named *units*, but trusted *standards* against which these lengths and weights and volumes and times and speeds, in whatever units they happened to be designated, could be measured.

The development of standards is necessarily very much more modern than the creation of units—and over the years there has been a steady evolution of the debates about standards, which in summary can be divided into three—whether they are and should be based on tangible human-scale entities—the thumb or the knuckle for the inch, say; or on created objects—man-made rods of brass or cylinders of platinum, say; or whether they should be based on absolute aspects of the natural world, carefully observed aspects which are immutable and constant and eternal.

It was Galileo who took the first step, in 1582, and by the simple act of noticing something quite mundane. It may or may not be legend: that while sitting in his pew in the cathedral at Pisa he watched the lantern over the nave swinging back and forth, and doing so at a regular rate. He experimented with a pendulum and found out that the rate of the swing depended not on the weight of the pendulum bob, but on the length of the pendulum itself. The longer the pendulum arm, the slower and more languid the back-and-forth interval. A short pendulum would result in a more rapid tick-tock, tick-tock. By way of Galileo's simple observation so length and time were seen to be linked—a linkage that made it possible that a length could be derived not simply from the dimensions of limbs and knuckles and strides, but by the hitherto quite unanticipated observation of the passage of time.

A century later an English divine, John Wilkins, proposed employing Galileo's discovery to create an entirely new fundamental unit, one that had nothing to do with the then-traditional standard in England, which was a rod that was more or less officially declared to be the length of a yard. In a paper published in 1668, Wilkins proposed quite simply making a pendulum that had a beat of exactly one second—and then, whatever the length of the pendulum arm that resulted would be the new unit. He took his concept further: a unit of volume could be created from this length; and a unit of mass could be made by filling the resulting volume with distilled water. All three of these new proposed units, of length, volume, and mass, could then be divided or multiplied by ten—a proposal which made the Reverend Wilkins, at least nominally, the inventor of the idea of a metric system. Sad to say, the committee set up to investigate the plan of this remarkable figure* never reported, and his proposal faded into oblivion.

Except that one aspect of the Wilkins proposal did resonate—albeit a century later—across the Channel in Paris, and with the support of the powerful cleric and

---

* Wilkins, who was variously warden of Wadham College, Oxford, and master of Trinity College, Cambridge, was a polymath the like of which is little known today. Not only was he a practicing priest and college administrator, friend to such as Christopher Wren (St. Paul's Cathedral) and Robert Boyle (of Boyle's law in physics), but he had a great interest in science: he suspected there might be life on the moon, imagined the existence of new planets, devised plans for submarines and aircraft and perpetual motion machines, and, in the same book that proposed a metric system based on the pendulum, proposed the establishment of a new universal language, because of the deficiencies of Latin. Also, during his time at Wadham, he created transparent beehives so that honey could be harvested without disturbing the bees.

diplomat Talleyrand. The formal proposal, which Tall-eyrand put to the National Assembly two years after the Revolution, in 1791, exactly duplicated Wilkins's ideas, refining them only to the extent that the one-second beating pendulum be suspended at a known location along the latitude of 45 degrees North. (Varying gravitational fields cause pendulums to behave in varying ways: sticking to one latitude would help mitigate that problem.)

But Talleyrand's proposal fell afoul of the postrevolutionary zeal of the times. The Republican Calendar had been introduced by some of the ardent firebrands of the day, and for a while France was gripped by a mad confusion of new-named months (Fructidor, Pluviôse, and Vendémiaire among them), ten-day weeks (beginning on *primidi* and ending on *décadi*), and ten-hour days—with each hour being divided into one hundred minutes and each minute into a hundred seconds. Since Talleyrand's proposed second did not match the Revolutionary Second (which was 13.6 percent shorter than a conventional second of the Ancien Régime) the National Assembly, gripped by the new orthodoxy, rejected the idea wholesale.

And it would be more than two further centuries before the fundamental importance of the second was fully accepted. For now, in the minds of eighteenth-century

French assemblymen, length was a concept vastly preferable to time.

For in dismissing Talleyrand so they turned instead to another idea, brand-new, which was linked to a natural aspect of the Earth, and so in their view more suitably revolutionary. Either the meridian of the Earth or its equator should be measured, they said, and divided into forty million equal parts, with each one of these parts being the new fundamental measure of length. After some vigorous debate, the parliamentarians opted for the meridian, in part because it passed through Paris; they then also decreed that to make the project manageable the meridian be measured not in its entirety, but only in the quarter of it that ran from the North Pole to the equator—a quarter of the way around, in other words. This quarter should then be divided into ten million parts—with the length of the fractional part then being named the meter (from the Greek noun μέτρον, a measure).

A great survey was promptly commissioned by the French parliament to determine the exact length of the chosen meridian—or a tenth part of it, an arc subtending about nine degrees (a tenth of the ninety degrees of a quarter-meridian), and which, using today's measurement, would be about a thousand kilometers long. It would necessarily be measured in the length units of eighteenth-century France: the *toise* (about six feet long), divided into six *pieds du roi*, each *pied* divided into twelve *pouces*, and these further divided into twelve *lignes*. But these units were of no consequence—because all that mattered was that the total length be known and then be divided by ten million—with whatever resulted becoming the measure that was now desired, a creation of France to be eventually gifted to the world.

The proposed survey line ran from Dunkirk in the north to Barcelona in the south, each port city self-evidently at sea level. Since this nine-odd-degree arc was located around the middle of the meridian—Dunkirk is at 51 degrees North and Barcelona 41 degrees North, with the midpoint of 45 degrees North being the village of Saint-Médard-de-Guizières in the Gironde—it was thought likely the oblate nature of the Earth's shape, the bulge that afflicts its sphericity and makes it resemble more of an orange than a football, would be most evident and so easier to counter with calculation. (To further confirm the Earth's shape the French Academy of Sciences sent out two more expeditions, one to Peru and the other to Lapland, to see how long a degree of high latitude was: all confirmed the orange shape that Isaac Newton had predicted centuries before.)

The story of the triangulation of the meridian in

France and Spain, and which was carried out by Pierre Méchain and Jean-Baptiste Delambre over six tumultuous years during the worst of the postrevolutionary terror, is the stuff of heroic adventure. On numerous occasions the pair escaped great violence (but not jail time) only by the skin of their teeth. The story is also outside the scope of this account, for what matters to precision engineers of the future—and to engineers all over the world, since that one remarkable survey led to the establishment of the metric system still in use today—is what the French did once the survey results were in. And that mostly involved the making of bronze or platinum rods.

The survey results were announced in April 1799. The length of the meridian quadrant was calculated from the extrapolated survey findings to be 5,130,740 *toise*. All that was required was that bars and rods be cut or cast that were one ten-millionth of that number—0.5130740 *toise*, in other words. And that length would be, henceforward, the standard measure—the standard meter—of postrevolutionary France.

The commissioners then ordered this length to be cast out of platinum, as what is known as an *étalon*—a standard. A former court goldsmith named Marc Étienne Janety had been selected to make it, and was called back from Marseille, where he had been sheltering from the excesses of the Terror. The result of his labors exists to this day—the Meter of the Archives, a bar of pure platinum that is twenty-five millimeters wide and four millimeters deep, and exactly, exactly, one meter in length. On June 22, 1799, this meter was officially presented to the National Assembly.

But that was not all: for in addition to the platinum rod that was the meter, so also there came with it a few months later a pure platinum cylinder which, it was explained, was the *étalon* of mass, the kilogram. Janety had made this one too, and also from platinum, 39 millimeters tall, thirty-nine millimeters in diameter, stored in a neat octagonal box with the label proclaiming, in good Napoleonic calendric detail, *"Kilogramme Conforme à la loi du 18 Germinal An 3, présenté le 4 Messidor An 7."*

The two properties of length and mass were now inextricably and ineradicably connected. For once the standard of length had been determined, so that length could be employed to determine a volume and, using a standard material to fill that volume, so a mass could be determined too.* And so in Paris at the exhausting end

---

* The linkage of length and mass standards, and the concept of using water to come up with a standard of mass, was first put forward by the same John Wilkins who suggested using a pendulum for length determination.

eighteenth century it was decided to create a new
ndard for mass based on a formula of elegant simplicity. One-tenth of the newly presented meter—and
which would be technically a decimeter—could be set
as the side of an exactly manufactured cube. This cubic
decimeter would be called a *litre* measure, and it would
be made as precisely as possible out of steel or silver. It
would then be filled entirely with pure distilled water
and the water held as close as possible to the temperature of 4 degrees Celsius, the temperature at which the
density of water is most stable. The resulting volume,
this one liter of this particular water, would then be
defined as having a mass of one kilogram.

The platinum object made by the goldsmith M. Janety was duly cast, and adjusted until it exactly balanced
the weight of that cubic decimeter of water. And that
platinum object—very much smaller than the water, of
course, since platinum was so much denser, by a factor of almost twenty-two—would from December 10,
1799, henceforward *be* the kilogram.

The Kilogram of the Archives and the Meter of the
Archives, from which the kilogram had been determined, were thus the new fundamentals of what would
soon be a new world order of weights and measures.
The metric system was now officially born.

These two icons of its founding are still in existence,

in a steel safe deep within the Archives Nationales de France in the Marais, in central Paris. One resides in an octagonal black leather-covered box, the other in a long and thin box of reddish-brown leather.

Except that—and this is a constant feature in the universe of measurement—these beauteous objects were eventually found to be wanting.

Years after they had been fashioned, the meridian line on which they had been based was resurveyed, and to widespread chagrin and dismay it was discovered that there were errors in Delambre and Méchain's six-year eighteenth-century survey, and that their calculation of the length of the meridian was off. Not by much, but by enough for the physical Meter of the Archives to be shown to be two-tenths of a millimeter shorter than the newly calculated version. And it follows that if the meter was wrong, then the cubic meter and the cubic decimeter and the liter-of-water equivalent in platinum, which would be the kilogram, would be wrong also.

So a cumbersome process was set in train to create a set of wholly new prototypes, which would be as perfect in their exactitude as late nineteenth-century science could manage. It took more than seven decades for the international community to agree, and many further years to make the requisite cache of bars and cylinders.

The mechanics of their making illustrates just how far the idea of precision had come in the century since John Wilkinson, boring his cylinders for James Watt, had come. The need to make the standards as near-perfect as imaginable was to become the stuff of obsession.

Fifty international delegates—all of them men, all of them white, and almost all of them with lengthy beards—gathered for the first meeting of the International Metre Commission in Paris in September 1872 to begin the process. They met in the former medieval priory of St. Martin des Champs, later to be turned into the Conservatoire National des Arts et Métiers, one of the world's greatest repositories of scientific instruments.*

The countries that would decide the future of the world's measurement system included all the then-great Western powers—Britain, the United States, Russia, Austria-Hungary, the Ottoman Empire—but pointedly, neither China nor Japan. Their sessions, and those of their associated conferences—most notably the Diplomatic Conference of the Metre, which was more

---

* One fewer after an accident in mid-May 2010 when the original Foucault's pendulum of 1851, housed in the Conservatoire for decades, crashed to the floor, irreparably damaging its bob. The cable had snapped; some said that attendees at private parties held in the museum were known to have played with the solemnly swinging pendulum, weakening its stays.

concerned with national policies, less with the technical aspects of making prototypes—went on for what at this remove seems an interminable period.

All of the meetings would, however, lead eventually to the signing, on May 20, 1875, of the Treaty of the Metre. It would mandate the formation of the BIPM, the present-day International Bureau of Weights and Measures, which would have as its home the Pavillon de Breteuil, outside Sèvres, and which it still inhabits today. Between them these bodies, at various times and in various ways, would commission the making of a set of vital new prototypes.

It took nearly fifteen years for the defining set of internationally agreed standard measures to be created, for the new standard artifacts to be cast, machined, milled, measured, polished, and offered up for the world's approval. On September 28, 1889, a ceremony was held in Paris to distribute them.

The two best made, each so perfect in their appearance and exact in their dimensions, and which in consequence were nominated to be the international prototypes, had by now been chosen: they were the International Prototype Meter, to be known hereafter by the black-type letter M, and the International Prototype Kilogram—*Le Grand K*—designated by the black letter K. Both of these platinum-iridium alloy objects were to remain for all future time under heavy security in the basement of the Pavillon de Breteuil.

All the others were then, and for this September day only, on display in the Pavillon's observatory. The stubby little kilograms gleamed under glass cloches (the national standards under a pair of glass cloches, the IPK itself under three), the slender meter bars in wooden tubes that were further enclosed in brass tubes with special fixtures to keep them safe while they traveled.

Certificates of authenticity had been engraved on heavy Japanese paper by the Parisian society printer Stern. Each of these certificates had a formulaic rubric that gave the properties of the body it accompanied: platinum-iridium cylinder No. 39, for example, had the notation "46.402mL 1kg - 0.118mg," which is decoded as meaning the cylinder had a volume of 46.402 milliliters and was lighter than 1 kilogram by 0.118 milligrams. Certificates for the meters were a little more complicated: for instance, one of the meter bars was noted as being "1m + 6 μ.0 + 8 μ.664T + 0 μ.00100T^2," which meant that at 0 degrees Celsius it was 6 micrometers longer than 1 meter, and at a 1 degree Celsius its length would be greater by a little more than 8.665 micrometers.

Three urns stood on a dais in the room, and officials had put into each paper slips bearing the numbers of the remaining standards—they were to be distributed among the member states by lottery. And so, in midafternoon of that warm autumn Saturday, the world lined up as if bidding for the distribution of sporting season tickets. Officials called out the countries' names, in alphabetical order, in French. Allemagne was first, Suisse last. The draw took an hour. When it was all over the United States had received Kilograms 4 and 20, and Meters 21 and 27.* Britain had acquired Meter 16 and Kilogram 18; Japan (which by this time had signed the 1875 treaty),† Meter 22 and Kilogram 6.

By the end of the day, so the delegates set off from Paris with their invaluable bounties—all packed away in boxes (the kilograms removed from their cloches for travel), and with all the bills paid. They were not insubstantial: the cost of a platinum-iridium meter was 10,151 francs; the kilogram a comparative steal at 3,105 francs. Within days or weeks (the Japanese took theirs back by

---

* After serving as the U.S. standard for the meter for seventy-one years—and being taken to Paris four times during that period, for comparison with *Le Grand K*—No. 27 was retired in 1960 and sits in a glass case in a museum at the National Institute of Standards and Technology in Gaithersburg, Maryland, outside Washington, DC.

† China would not be a party to the treaty until 1977—by which time, as we shall see, the entire system of measurement had changed.

ship) the new standards were safely in the metrology institutes that were by now being established in capitals all around the world. They were all kept safe and sound—though none so safe and sound as the International Prototypes, M and K, which were now to be taken to the basement and plunged into sempiternal darkness, incomparable, accurate, and fantastically precise. In safes nearby were six so-called *témoins*—witness bars, which would be regularly compared against the masters. These too would remain exact and perpetually inviolate.

Except, not exactly. Not so fast. The overseers of metrology's fundamentals had been charged with the task of eternal vigilance, of always looking for still better standards than these. And in time they did indeed find one.

The first clues that there might be a better system had come some years before, in 1870, long before these platinum talismans were being wrought into their final definitive shapes and sizes. The Scots physicist James Clerk Maxwell, at the British Association for the Advancement of Science annual meeting in Liverpool, had made a speech that threw a wrench into everything that had been done. His words still ring in the ears of metrologists around the world. He reminded his listen-

ers that modern measuring had begun with the survey
and then the resurvey of the French meridian, and the
derivation of the metric units from the results:

> *Yet, after all, the dimensions of our Earth and its*
> *time of rotation, though, relatively to our present*
> *means of comparison, [are] very permanent, [they]*
> *are not so by physical necessity. The Earth might*
> *contract by cooling, or it might be enlarged by a*
> *layer of meteorites falling on it, or its rate of revolu-*
> *tion might slowly slacken, and yet it would continue*
> *to be as much a planet as before. But a molecule,*
> *say, of hydrogen, if either its mass or its time of*
> *vibration were to be altered in the least, would no*
> *longer be a molecule of hydrogen.*
>
> *If, then, we wish to obtain standards of length,*
> *time and mass which shall be absolutely perma-*
> *nent, we must seek them not in the dimensions, or*
> *the motion, or the mass of our planet, but in the*
> *wavelength, the period of vibration, and the abso-*
> *lute mass of these imperishable and unalterable and*
> *perfectly similar molecules.*

What Maxwell had done was challenge the scientific basis for all systems of measurement up to that
moment. It had long been self-evident that a system

based on the dimensions of the human body—thumbs, arms, stride, and so forth—was essentially unreliable, subjective, variable, and useless. Now Maxwell was suggesting that standards previously assumed reliable, like fractions of a quadrant of the Earth's meridian, or the swing of a pendulum or the length of a day, were not necessarily usefully constant either. The only true constants in nature, he declared, were to be found on a fundamental, atomic level.

And by this time scientific progress was providing windows into that atom, revealing structures and properties hitherto undreamed of. These very structures and properties that were by their very nature truly and eternally unvarying, Maxwell was saying, should next be employed as standards against which all else should be measured. To do otherwise was simply illogical. Fundamental nature possessed the finest standards—the only standards, in fact—so why not employ them?

It was the wavelength of light that was the atomic fundamental first used to try to define the standard measure of length, the meter. Light, after all, is a visible form of radiation caused by the excitation of atoms—excitation that causes their electrons to jump down from one energy state to another. Different atoms produce light ranging over different spectrums, with different wavelengths and colors, and so produce different and identifiable lines on a spectrometer.

It took a further hundred years to convince the international community of the wisdom of linking length to light and its wavelength. To the graybeards who then ran the world, abandoning the certitudes of Earth for the behavior of light was akin to believing that the continents could move—a simply preposterous idea. But just as in 1965, when the theory of plate tectonics was first advanced and continental drift was suddenly seen as obvious, a reality hidden in plain sight, so it became as much the same in metrology as it had been for geology: the notion of using atoms and the wavelength of the light they can emit as a standard for measuring everything snapped into place in a sudden moment of rational realization.

It was a late nineteenth-century Massachusetts genius named Charles Sanders Peirce who had that first moment, who first tied the two together. Few men of his generation can have been more brilliant—or more infuriatingly, insanely troublesome. He was many things—a mathematician, a philosopher, a surveyor, a logician, a philanderer of heroic proportions, and a man crippled with pain (a facial nerve problem), with mental illnesses (severe bipolar disorder most probably), and with a profound inability to keep his temper in check.

On the plus side of the ledger: he could stand before a blackboard and write a mathematical theory on it with his right hand on the right side and, simultaneously, write its solution with his left hand on the left. On the minus side: he was once sued by his cook for hitting her with a brick. He drank. He took laudanum. He was much married, and was pathologically unfaithful.

But it was Peirce who in 1877 first took a pure and brilliant source of incandescent yellow sodium light, and tried as hard as he might to measure—in meters, thereby establishing the dimensional link between light and length—the black spectral line it produced when run through a diffraction grating, a kind of high-precision prism. It was one of the numberless misfortunes of his seventy-five years that this experiment never quite succeeded—there were problems with the expansion of the glass of the grating, problems with the thermometers used to measure the temperature of the glass. But he nevertheless published a short paper in the *American Journal of Science*, and by doing so laid historical claim to being the first to try. Had he succeeded his name would be on the lips of all. As it was he died obscurely in 1914, and in abject poverty, having to beg stale bread from the local bakery. He is long forgotten, except by a very few who agree with such as Bertrand Russell, who called Peirce "the greatest American thinker, ever."

By 1927, after much badgering by scientists who were convinced by Maxwell's argument that this was the best approach to setting an inviolable standard, so the world's weights and measures community came, if somewhat grumpily, to an agreement. They first accepted, formally, that one particular element's wavelength had thus been calculated, and in fractions of a meter—a very small number. Further, they then agreed that by multiplication, the meter could be defined as a certain number of those wavelengths—by comparison a very big number, and to at least seven decimal places. Multiply the one by the other and one gets, essentially, one meter.

The element in question was cadmium—a bluish, silvery, and quite poisonous zinc-like metal that was used for a while (with nickel) in batteries and to corrosion-proof steel and now is used to make (with tellurium) solar panels. It emits a very pure red light when heated, and from its spectral line the wavelength could be determined—so accurately that the International Astronomical Union used its wavelength to define a new and very tiny unit of length, the Ångstrom—one ten-billionth of a meter, 10^{-10}m.

The wavelength of cadmium's red line was measured and defined as 6,438.46963 Ångstroms. Twenty years later, with the weights and measures officials in

Paris now accepting both the principle and the choice of cadmium (although making its red-line wavelength slightly fuzzier by losing the final number 3, rendering it as 6,438.4696Å), the meter could have been very easily defined by simple arithmetic as 1,553,164 of those wavelengths. (Multiplying the first figure by the second gives 1.000, essentially.)

But—and in the tortuous history of the meter, this is hardly surprising—cadmium then turned out to be not quite good enough. Its spectral line, when examined closely, was found not to be as fine and pure as had been thought—the samples of cadmium were probably mixtures of different isotopes of the metal, spoiling the hoped-for coherence of the emitted light. And so it happens that the meter never was formally defined in terms of cadmium. Much else was, but not the sacrosanct meter. The platinum-iridium bar clung on gamely through all the various meetings of the weights and measures committees, surviving all the siren-like temptations of other radiations—until finally, in 1960, there came agreement.

The world settled on krypton. This inert gas, which was only discovered in trace amounts in the air in 1898, is perhaps best known as the most commonly used gas in neon signs, which are seldom filled with neon at all. More important, in this long quest to define the meter in terms of wavelength, krypton has a spectral signature with extremely sharp emission lines. Krypton-86 is one of the six stable isotopes that occur naturally,* and on October 14, 1960, the International Committee on Weights and Measures decided, nearly unanimously, that this gas, with its formidable coherence and with the exactly known wavelength of its emissions of reddish-orange radiation (6,057.80211Å) would be the ideal candidate to do for the meter what cadmium had done for the Ångstrom.

And so, with the delegates observing that the meter was still not defined with "sufficient precision for the needs of today's metrology," it was agreed that henceforward the meter would be defined as "the length equal to 1,650,763.73 wavelengths in vacuum of the radiation corresponding to the transition between the levels 2p10 and 5d5 of the krypton-86 atom."

And with that simple declarative sentence so the old one-meter platinum bar was pronounced, essentially, useless. It had lived since 1889 as the ultimate standard for all length measurement: Ludwig Wittgenstein had once observed, with confusing but accurate drollery,

---

* The unstable isotope krypton-85, which has a half-life of eleven years, is a by-product of nuclear explosions and fuel reprocessing—and the presence of plumes of the gas in the upper atmosphere has been detected by satellites orbiting over North Korea.

"There is one thing of which one can say neither that it is one meter long, nor that it is not one meter long, and that is the standard meter in Paris." No longer, for from October 14, 1960, onward, there was no standard meter remaining in Paris, nor anywhere else. This measurement had left the physical world and entered the absolutism and indifference of the universe.

Much else besides went on at the 1960 conference, which is held every four peacetime years, and usually in Paris, which on this occasion was perhaps the seminal event in metrology since the invention of the science. Most memorably, the 1960 event saw the formal launch of the present-day International System of Units, known generally by SI, initials derived from the French Système International d'Unités. Most of the world now knows, accepts, recognizes, and uses the SI—with its seven units: of length (the much-aforementioned meter); of time (the second); of electric current (the ampere); of temperature (the kelvin);* of light intensity

---

* Quite reasonably there is little that passes for poetry in any of the definitions, though maybe some will spy a hint of romance in the kelvin, which is defined as 1/273.16th of the temperature of the triple point—when liquid, solid, and vapor all coexist—of water. But not just any old water: the definition requires the use of what is known as Vienna Standard Mean Ocean Water, a cock-

(the candela); of the amount of a substance (the mole); and of mass (the kilogram). Six of these units are now defined in terms of natural phenomena—generally, of radiation and the behavior of or the number of atoms.

So much else came out of the meeting: the base units; the derived units—like the hertz, the volt, the farad, ohm, lumen, becquerel, henry, coulomb; the authorized prefixes for big and small—at the upper end the deca, kilo, giga, tera, exa, zetta, and yotta (this last being 10^{24}) and at the lower deci, milli, nano, pico, femto, zepto, and yocto (this, to preserve metrologic symmetry, denoting the phenomenally tiny 10^{-24}).

But what did not come out of the meeting was anything definite to relieve the condition of the other old standard, *Le Grand K*. The delegates—who had created an entirely new measurement system, after all—left Paris that late October, leaving behind them, condemned to remain locked up in a dark cellar under its triple crowns of glass, the melancholy standard mass of the kilogram, moping, miserable, a relic of an earlier century. It would take almost another sixty years for them to find a replacement, and for the highly polished

tail of various distilled waters drawn from all the oceans—and yet perversely named for the capital city of a landlocked country about as far from the sea as it is possible for any part of Europe to be.

solid metal cylinder, about as tall and wide as a Zippo lighter, about the size of a golf ball, to be relieved of its responsibilities of being the mass against which all the world's kilograms could and would be measured: in late 2018, it is to be removed from under its well-guarded basement cloche and placed in a museum—a relic of former times, of more ancient technologies.

And since the kilogram's replacement was to occur so much later than that of the meter, so it enjoyed the benefits of metrology's even newer technological evolution. For it was to become related to a unit that had long been overlooked as the key to all others—and that is the unit of time, the second.

It has to do with the notion of frequency, which is after all the inverse of time—it is the number of occurrences of something *per second*. And frequency is now mentioned in no fewer than six of today's seven foundational units of measurement.* Frequency is just about everywhere.

---

* As mentioned before, the seven fundamental units are the kilogram (of mass); the meter (of length); the second (of time); the ampere (of electrical current); the kelvin (of temperature); the candela (of luminous intensity); and the mole (of molecular amount). There are a range of what are called "derived units"

Three examples will suffice.

The candela, the unit that suggests the brightness of a source of light, would seem at first blush to have absolutely nothing to do with time. But it has: the international community now defines the candela as *the luminous intensity, in a given direction, of a source that emits monochromatic radiation of frequency 540 × 10^{12} cycles per second and that has a radiant intensity in that direction of 1/683 watt per steradian.* Light is here officially related to the second. It is officially linked to the concept of time.

The length of the meter, to select another of the seven units as an example, is now also defined in terms of the second—*it is the length of the path traveled by light in vacuum during a time interval of 1/299,792,458 of a second.* Length, henceforward (or since 1983, when it was so defined), is thus also related to time. A relationship that is agreed to by all.

---

to supplement these—such as the coulomb (electrical charge); the newton (force); the pascal (pressure); the farad (electrical capacitance); and some fifteen more, including the popular tesla, which, though defining an obscure property called the "magnetic flux density," memorializes one of science's most popular recent scientists, Nikola Tesla. He won this honor in 1960, seventeen years after his death.

And the much-vaunted kilogram, until lately defined as the carefully milled platinum cylinder in Paris, will soon reappear, defined this time in terms of the speed of light—and connected to it by way of the famous Planck constant, which, without going into the details of the thing, is a number, $6.62607004 \times 10^{-34}$ m^2 $\times$ kg/s, that, as the symbols imply, is also firmly linked to frequency, and thus to the second. Mass is thus defined in terms of time. The whole world now agrees it should be so: that time underpins everything.

Just as Galileo had so presciently realized when he gazed up at the lantern in Pisa. Just as Wilkins had later proposed, and the Prince of Talleyrand seconded. All are connected by time.

And yet—just what is time?

"If no one asks me," Saint Augustine is said to have remarked, "I know what it is. If I wish to explain it to him who asks, I do not know." Time moves, we know that. But how does it move? What is its moving, exactly? And why does it only move forward, in one direction? And so far as time is concerned, what does direction mean, exactly? Can one be any more precise than simply to say, as Einstein once did, that time *is what clocks measure?*

All such questions are suddenly especially pertinent.

How we arrange—and how in history we have arranged—the accumulations of time is a matter of choice. On the matters of minutes and hours and days most generally agree*—after all, the sun's rising and setting have long dictated the nature of time, creating a top-down arrangement that was made for the convenience of human society, and allowed for the notion, even as recently as the 1950s when the second, at the bottom of this top-down arrangement, was defined as 1/86,400th of the passage of a single day.

Beyond days—up into the other human constructs known in English as *weeks* and *months* and *years*—the arrangements became wildly different according to the vagaries of religion and custom and the caprices of per-

---

* The sixty-second minute, the sixty-minute hour, and the twenty-four-hour day may now be near-universally accepted, but France has a long tradition of preferring decimal time, which supporters insist is logically connected to the decimal divisions of length and mass. For many centuries China divided its days in decimal fashion, but did so somewhat capriciously—there were extended periods of Chinese history when the basic unit, the *ke*, differed markedly in its duration from other periods. In the seventeenth century, the Jesuits brought harmony to Chinese timekeeping, declared the *ke* to be a quarter of an hour, and thence gently shepherded China into diurnal conformity with the rest of the world.

sonalities. But it is the considered aim of modern metrologists, when dealing with matters concerning the
basic unit, the second, that all of the units should agree,
exactly. So far as larger units of time are concerned,
all are free to do as they wish. But the second itself is
sacrosanct.

Until 1967 the second was very much linked to a
natural phenomenon—as the fraction of the length of
the day, at the top of the top-down pyramid—by way
of a sundial or by a seconds pendulum, which ticked
away the duration of a day at intervals that were determined by the length of the pendulum itself. It was easy
enough—if time consuming—to adjust the length of a
pendulum until it ticked away at the rate of 1/86,400 of
the period between two sun-at-zenith moments we call
noon. Easier still to apply the equation from schooldays,
of $T = 2\pi \sqrt{lg}$, where l is the length of the pendulum, g
is the acceleration of gravity, and T is the time taken by
each beat of the pendulum.

To deduce the second from the day is indeed easy
enough. The greater problem, recognized from antiquity, is that the length of the day itself turned out to be
almost infinitely variable, due to a range of reasons both
local—such as the frictional effects of the tides—and
astronomical—such as changes in the Earth's rotation,
the wobbling-top precession of its axis, the steady slow-

ing (and occasional random speeding-up) of the Earth's period of rotation. For how can a second be accurately defined if the standard against which it is measured is inherently unstable? This was James Clerk Maxwell's singular problem, once again.

The way that this problem was first dealt with was to replace the day at the top of the notional pyramid with the much larger unit of the year—and to measure increments of time as fractions of the year, of the passage of time taken for the Earth to make one complete turn around the sun. The notion of *ephemeris time* was born with this decision—ephemeris time being based on the movements of the planets and the stars as recorded from centuries of observation.

Tables known as ephemerides—*almanacs* a less confusing term—get better and better as the years go on, because of the ever-more- sophisticated observations first from telescopes and, later on, from satellites. And so, the modern concept of ephemeris time, defined by the Jet Propulsion Laboratory in Pasadena, became standard, in 1952.

A second was then defined as $1/31,556,925.975$ of a year—and not just any year, but the year 1900 beginning on January 0—this last being a way of using the midnight handover of 1899 December 31 to 1900 January 1 as the starting point, and ignoring the fact that, inconveniently for some, years—being a human construct—never begin with a day labeled as 0. Our counting system does (0.5); our clocks do (00.23 h); but our calendars (January 1, never January 0) do not.

But then the year itself, based as it is on the wanderings of a planet around a star, was found to be just as arbitrary and as wanting in precision as was the day, and so better still was needed. As it happens a better solution was waiting in the wings: Maxwell's answer. That there are things in nature, most especially in atomic and subatomic nature, that vibrate at frequencies which never, ever, ever change. Or not to any measurable degree.

Quartz, as we discovered back at Seiko, is one such. The seconds presented by a quartz-based timekeeper were unvaryingly precise seconds; and the seconds they soundlessly accumulated turned into precise minutes, precise hours, precise days.

And yet, just as with Maxwell's argument against using a human-scale or even a planetary-scale basis for defining the meter and the kilogram, so in the latter half of the twentieth century it became clear that though quartz is good enough for the average consumer of time, it is manifestly not good enough for the scientist, nor for the national metrology institutes around the world. Which led to the evolution of the standards that are in use today, and which employ one or more members of the more recently invented families of *atomic clocks*.

In an atomic timekeeper the same basic principle applies—that a naturally occurring substance can be induced to vibrate at a certain fixed and measurable rate. With a quartz crystal, it was the simple and easily knowable property of its vibration under the influence of an electrical charge that made it so attractive a candidate for timekeeping. With an atom, the frequency was a more delicate thing: it required that an electron in orbit around the nucleus of a candidate element be persuaded to shift to another orbit—to make a quantum leap, or a quantum jump, this being the origin of the phrase. It had been known since the nineteenth century that when an electron performs this leap from its ground state to another energy level, it emits a highly stable burp of electromagnetic radiation.

The radiation from such an atomic transition, it was said by many physicists, was so exact and so stable that it might well one day be used as the basis of a clock. The basic concept was first demonstrated in the United States in 1949, in a precursor to the laser, the maser, and which employed molecules of ammonia.

The first true atomic clock was invented by a Briton, Louis Essen, in 1955, when he and a colleague, Jack Parry, made a model and used as its heartbeat the transition of electrons orbiting the nucleus of atoms of the metal cesium. This might seem a curious choice: cesium is the softest of all metals—almost liquid at room temperature—and is a pale gold-colored substance that ignites spontaneously in the air and explodes when in contact with water. However, it now has a use and value beyond all measure, since in transition it emits radiation at such a steady and unvarying beat that the scientists at Sèvres readily agreed, in 1967, and after much badgering by Louis Essen and Britain's National Physical Laboratory, where he worked, that it be used as the basis for a new definition of the second.

As it remains today. The definition of the second today is quite simply, if simply be the word, *the duration of 9,192,631,770 cycles at the microwave frequency of the spectral line corresponding to the transition between two hyperfine energy levels of the ground state of cesium 133.* The ten-digit number, daunting though it may sound at first, is known by every metrologist worth his or her salt, and is as familiarly and frequently bandied about as might be an American telephone number, and which it digitally resembles.

Cesium clocks are now everywhere, costly and bulky though they still may be. There are said to be 320 of them, all checked against one another—the American master clocks checked every twelve minutes to eradicate nanosecond errors. All these are then checked themselves by squadrons of even more accurate time-keepers called cesium fountain clocks, of which there are a dozen, and which employ lasers to roil a mess of cesium atoms inside a steel vessel and derive even greater accuracies than their simpler siblings.

In America, the master clocks are in Maryland and Colorado; and the GPS system—the highly precise and time-based creation described in chapter 8—is given its critical time data from an ensemble of no fewer than fifty-seven cesium clocks held at the U.S. Naval Observatory* in Washington, DC, and which in turn are augmented by a further twenty-four at the formidably well-protected Schriever Air Force Base in Colorado.

The accuracy of these clocks and the claimed accuracies of even newer ones that are being constructed or experimented with at various standards laboratories around the world—the ytterbium clock being studied

---

* The USNO was built on a low hill near the British embassy on Massachusetts Avenue, the site chosen to avoid light pollution from the then-small city to its south. Now it is surrounded by suburbs and a consequent vast amount of emitted light—and also as it happens by regiments of Secret Service guards, there to protect the onetime Superintendent's Mansion, now the official residence (with a nuclear-hardened bunker beneath) of the American vice president.

at the National Institute of Standards and Technology outside Gaithersburg, Maryland, being a prime example—begin to verge on the barely credible. The British Standards Institution, for example, has claimed that while the standard cesium clock has an accuracy to some 10^{-13} seconds, with its fine-tuned cesium fountain clock known as NPL-CsF2, the second could be measured to a known degree of precision of 2.3×10^{-16}, or 0.000 000 000 000 000 23.

This means it would neither lose nor gain a second in 138 million years.

Now there is talk of quantum logic clocks and optical clocks that deliver even more remarkable figures, one with a claimed accuracy of 8.6×10^{-18}, meaning that time would be kept impeccably for *a billion years*, and the charming concept of taking the fob watch from the pocket every few days and lovingly adjusting it would be gone forever, both from the human imagination and from memory.

It is into this rarefied world of precise chronometry that science has now jumped—pouring money and equipment and personnel into matters relating specifically to the measurement of the bizarreries of time—and for the simple reason, fully recognized by teams of metrologists, that time *underpins everything.* "Everything" even includes, it now seems, the property of gravity. A clock that is on a table just five centimeters higher than another will record seconds that are barely measurably longer—but incontrovertibly longer, nonetheless—than its partner. And this is simply because it is less affected by the Earth's gravity, the planet's center being that tiny number of centimeters more distant.

This link, between time and gravity, is now proven. And this is a happenstance of modern physics that in China—where much work is being conducted on the nature of time—has a certain unanticipated charm. There is a certain delight of synchrony for the metrologists who are conducting time-related experiments in their brand-new and handsomely funded laboratories near Beijing. For outside the very front door of their research center there stands a gift from England's main metrology institute, the National Physical Laboratory in Teddington, west of London.

It is a sapling apple tree.

Outwardly it looks quite ordinary—just one tree among a copse of others. But this happens to be a very special tree indeed. If the Beijing summers are warm and not too dry, it will bear apples of the variety known as Flower of Kent, which are said to be crunchy, juicy, and acidic. But this is not the reason. It is the tree's pedigree that marks it out as unique.

Before the NPL gave it as a gift the apple tree's immediate ancestor was grown from a scion that had been grafted in the 1940s at a fruit-research station south of London, which in turn had been taken from a tree in the garden of an abbey in Buckinghamshire, and which had been planted in the 1820s. This in turn was a relic of a mighty tree that had been blown down in a great historic storm that had devastated a country estate a little farther north, that of Woolsthorpe Manor in Lincolnshire.

And Woolsthorpe Manor was the home of Sir Isaac Newton. It was to Lincolnshire that Newton had fled from Cambridge in 1666—and it was here, during the summer of that *annus mirabilis*, that he famously observed the apple falling from the tree. It was here, and from wondering of the force that might have impelled the apple's fall, that he came up with the notion of gravity, as a force that affected both this humble fruit and by logical extension affected the constant motion and altitude of the moon in orbit around the planet Earth.

So, Isaac Newton's apple tree—or more properly a child descendant of it—now flowers and fruits in a Beijing garden, beside where the Ming emperors once buried their dead, where one can see the Great Wall running along the mountain ridges, and where China's latest generation of scientists are confirming their intellectual ambitions by working out, with the greatest accuracy, the effect that gravity has upon the steady beat of time.

Where, in other words, they are trying to establish and prove a physical, traceable connection between on the one hand the mysterious force that keeps us all rooted here on Earth, and on the other the fundamental steady tick of duration. The duration by which, fundamentally, we measure everything that we make and use, and which in turn helps establish for us with unfailing exactitude the precision that allows the modern world to function.

Acknowledgments

For at least the last seven centuries the decorative brass faceplate of an astrolabe has been known as a "rete." The word came into the English language from the Latin for "network," and in this lexical sense it very much works, since the face of many an old astrolabe looks much like a metallic net which has been cast over the more solid wheels and gears that make up this most ancient of astronomical instruments.

The word is also employed nowadays on the internet. It denotes a mail-list, a perpetual cyber-conversation run from the Museum of the History of Science in Oxford, and which connects in an immense electronic network a worldwide group of people who are fascinated by the intersecting topics of measurement, scientific devices (astrolabes and orreries included, of course), as well as optics, cypher machines, and the dueling concepts of accuracy and precision. I joined this list back in 2016, doing so with a timid query to the effect that I was wanting to write a history of precision, and did anyone out there have any ideas?

Zounds! I was promptly flooded by a crashing wave of enthusiasm from all around the planet, from Potsdam to Padua, Puerto Rico to Pakistan, with legions of scientifically minded people offering me advice and sending me books, giving me links to academic papers, invitations to conferences and, by the score, the names of leading figures in the world of exactitudinal studies.

And so my first order of business is to thank the originators and organizers of the rete mail-list, and to honor the helpful multitudes of those who like to be known as "retians," for getting me started. A goodly number of the names that follow were aficionados whom I first encountered through rete@maillist.ox.ac.uk, and all were eager to be helpful in ways both great and small. Among them were:

Silke Ackermann, Chuck Alicandro, Paul Bertorelli, Harish Bhaskaran, John Briggs, Stuart Davidson, Michael dePodesta, Cheri Dragos-Pritchard, Bart Fried, Melissa Grafe, Siegfried Hecker, Ben Hughes, David Keller, John Lavieri, Andrew Lewis, Mark McEachern, Rory McEvoy, Graham Machin, Diana Muir, David Pantalony, Lindsey Pappas, Ian Robinson, David Rooney, Christoph Roser, Brigitte Ruthman, James Salsbury, Douglas So, Peter Sokolowski, Konrad Steffen, Martin Storey, William Tobin, James Utterback, Dan Veal, Scott Walker.

Many of these, and hosts of others, swiftly insisted after my first inquiry that I make contact with the two leading experts in the field of precision, Pat McKeown of Cranfield University in southern England, and Chris Evans of the University of North Carolina at Charlotte. I traveled to see both, and each proved a fountain of assistance and generosity. This book could hardly have been written without their help and encouragement, and my debt to both is considerable. Any errors or infelicities are, of course, mine alone.

During my research I visited the national metrology institutes in Britain, Japan, China, and the United States, and so wish to record my particular thanks to Paul Shore, Laura Childs, and Sam Gresham at the National Physical Laboratory in Teddington; to Gail Porter, Chris Oates, and Joseph Tan at the National Institute of Standards and Technology in Gaithersburg; to Kelly Yan at the Changping campus of the National Institute of Metrology in Beijing; and to Toshiaki Asakai and Kazuhiro Shimaoka at the National Metrology Institute of Japan in Tsukuba, as well as offer my gratitude for the valuable advice of Professor Masanori Kunieda at the University of Tokyo.

NASA scientists and other colleagues involved in both the Hubble and James Webb Space Telescopes were most helpful, including Mark Clampin and Lee Feinberg at the Goddard Space Flight Center, as well as Eric Chaisson at Harvard and Matt Mountain at AURA.

I wish also to give particular thanks to Richard Wray, Chloe Walters, and Bill O'Sullivan at Rolls-Royce in Derby; to Billi Carey of the Rolls-Royce Silver Ghost Society; to Mark Johnson, Andrew Nahum, Ben Russell, Jim Bennett, and Jenni Fewery at the Science Museum, London; to Jelm Franse at ASML in Eindhoven (as well as to my old friend Toni Tack for offering me hospitality and shelter while in the Netherlands during this expedition); to John Grotzinger and Ed Stolper at Caltech; to Steve Hindle at the Huntington Library in Pasadena (where I was briefly a scholar-in-very-comfortable-residence); to Richard Ovendon, who is Bodley's Librarian (and occupant of what must be one of the world's nicest offices) in Oxford; to Fred Raab and Michael Landry at LIGO, Hanford; to Jessica Brown of Northrop Grumman; to Keiko Naruse and Takashi Ueda of Seiko; to Stefan Daniel of Leica—and my old newspaper colleague Chris Angeloglou, a formidable Leica collector.

Stephen Wolfram and his colleague Amy Young, both immensely knowledgeable about precise measurement, provided wise counsel (and in Amy's case, a gift of Christmas cookies). Jeremy Bernstein, expert on all things nuclear, told me much about plutonium. Max Whitby, a friend of forty years, offered fascinating insights into the world of nanotechnology. And the Master of my old Oxford college, Roger Ainsworth, turned out to have been a lead member of the Rolls-Royce Blade Cooling Research Group, and offered valuable reminiscences. Ann Lawless of the American Precision Museum, in Windsor, Vermont, was a supporter of this book from its inception. The writer Witold Rybczynski and the filmmaker Nathaniel Kahn also expressed interest and encouragement, for which I am most grateful.

My son, Rupert Winchester, an always keen and close reader, provided invaluable commentary on the near-finished script, as he has for nearly all of my books.

I cannot speak too highly, nor too warmly, of my new editor at HarperCollins, Sara Nelson, who brought her years of expertise to bear on the initial manuscript and turned it into a finished document of which I am now cautiously proud. It has been the greatest pleasure to work with her, and we have achieved a rapport, I believe, which will now endure for many years to come. Her assistant Daniel Vazquez, succeeded in the closing weeks of this book's birth-canal progress by Mary Gaule, both proved more than worthy of Sara's confidence in them, and proved a delight to work with. Likewise, across in London, my HarperCollins editor, Arabella Pike: we are new to each other—but the fact that she bought for her young son a set of Jo blocks as a Christmas gift, and after reading about them in these pages, suggests that a long and happy friendship is in the making.

My thanks as always to my agents at William Morris Endeavor, Suzanne Gluck in New York and Simon Trewin in London, as well as to Andrea Blatt, Suzanne's assistant. Your tenacity and persistence are deservedly the stuff of legend, and as a beneficiary I am hugely grateful. For that, of course, but on a higher level for your friendship too, and which I know will long endure.

And finally, to my wife, Setsuko, both for offering her highly original insights into the fugitive relationship between precision and craft—most especially in Japan—and for her overall enthusiastic support for this book. My gratitude to her is boundless.

Simon Winchester
Sandisfield, Massachusetts
March 2018

A Glossary of Possibly Unfamiliar Terms

ACCURACY: The closeness of a measurement, or of an action, to a desired result. Hitting a bull's-eye displays accuracy.

ARMILLARY SPHERE: An elaborate framework of intersecting brass rings created to represent the various astronomical and other features—the ecliptic, say, or the lunar orbit or the tropics—around planet Earth.

ASTIGMATISM: Distortion in vision, or in the function of a camera or telescope, caused by refractive irregularities in the shape of a lens.

ASTRARIUM: A mechanical device, with similarities to both a planetarium and an astronomical clock, which can predict celestial events and planetary passages across the sky.

ASTROLABE: A graduated metal disc with rotating parts used to calculate astronomical events.

AVOGADRO NUMBER: The number of particles—atoms, photons, molecules—contained in the amount of a substance given by the SI unit of one mole. The number is expressed as 6.0221415×10^{23}.

BIMETALLIC STRIP: Given that different metals behave differently at different temperatures, joining small pieces of two such metals together will allow the resulting strip to bend as the temperature changes—a phenomenon used to trip switches in, say, a thermostat.

BLOCK (MARITIME): A pulley or system of pulleys mounted inside a case and used to hoist the rigging of a sailing ship, and to raise great weights.

BRAZING: A means of employing heat to join metal to metal, the result less permanent than welding, more so than soldering.

BREVET D'INVENTION: A French patent.

BRITISH STANDARD WHITWORTH: A set of standards for the making of screws and screw threads.

CALIPER: The jaws of a measuring instrument, one fixed and the other either pivoted or sliding, used to determine the outer dimensions of an object.

CARBINE: A gun, slightly shorter and lighter than a musket, and generally carried by mounted cavalry.

CARBON FIBER: A phenomenally strong and light form of carbon first made in the 1960s and enjoying all kinds of uses in mechanical construction.

CHAMFER: To smooth away a sharp edge or corner.

COMA (IN TELESCOPES): A result of spherical aberration in a lens, when objects are surrounded with blur.

COMPLINE: The final religious service of a monastic day, and by extension, the evening hour when it is held.

DIFFRACTION GRATING: A plate ruled with very fine parallel and equidistant lines which produces a vivid spectrum when light is passed through them.

DIVIDING ENGINE: A device, usually a large wheel advanced by a worm gear, used to incise graduations onto a measuring instrument.

DOWEL: A rod, or peg, usually smooth, often made of wood, used as a fastener.

EDM: Electrical discharge machining, in which high-intensity sparks are used to create shapes in metal workpieces—often used to drill tiny holes in inconveniently shaped items.

ETHER: An invisible substance formerly believed to fill all space and which permits the carriage of radiation of all kinds.

FAB: A fabrication plant for the bulk manufacture of electronic components.

FLINTLOCK: The device that more or less reliably produces, by striking a metal piece against a flint, the spark that will ignite the powder in a gun.

FOUCAULT'S PENDULUM: Named for the nineteenth-century French physicist who discovered the phenomenon: a very long and slow-moving pendulum which will keep oscillating in one direction while the Earth can be shown—by means of a dial under the bob—to be rotating beneath it.

FRIZZLE: The metal part of a flintlock (q.v.) that, when struck against the flint, produces a spark.

GLIDE PATH: The most prudent line of descent for an aircraft about to make a landing.

GO AND NO-GO GAUGE: An inspection tool, with two tolerance options—one a part (a rod, or a screw) into which the piece to be inspected will fit, and the other that will not fit, and from which one can then determine the dimensional correctness of the piece.

GOVERNOR: A mechanical device attached to an engine which can regulate and limit the engine's speed.

GRAPHENE: A form of carbon, produced in barely-visible sheets one molecule thick, artificially created in 2004 and now widely studied because of its formidable strength and lightness.

HA-HA: An artificial ditch created, often in large estates, as a near-invisible boundary around fields, meadows, and gardens.

INTERCHANGEABLE PARTS: A basis for modern manufacturing whereby all the component parts are made to be identical one with the other, so they will always fit when assembled. The concept, first developed in eighteenth-century France, dominated the American manufacturing industry of the nineteenth century.

INTERFEROMETER: A highly accurate optically based measuring device which splits a beam of light in two and recombines it such that differences in length along the paths followed by the beams cause the light waves to interfere and produce colored patterns from which differences in length can be deduced.

JIG: A guide, or support, often made by hand, and which helps position a tool or drill so that an operation can be repeated many times.

LATHE: A turning machine, its origin of great antiquity, in which the piece to be turned—usually wood or ivory or metal—is firmly held in a horizontal position and rotated while tools are brought to bear upon it.

LIGNUM VITAE: Usually refers to the wood of a West Indian tree, known for its exceptional hardness and self-lubricating properties. It was long used to make gearwheels and other mechanical parts. It will sink in water.

LIGO: The Laser Interferometer Gravitational-Wave Observatory, though usually referring to the two American-based observatories in Louisiana and Washington, can extend to a worldwide network of devices recording and measuring the passage of gravitational waves through the fabric of space-time.

MACHINE TOOL: A usually nonportable tool used for drilling or milling or shaping metal; often said to be a machine that makes machines.

MAGNETO: A small device that, employing magnets and coils of conducting wire, will produce a spark when turned mechanically.

MATINS: The first of the daily religious rituals of a monastic day, of which compline (q.v.) is the last.

MICROMETER: A measuring device, often based on a finely made screw and a graduated means of turning it to advance and retard its parts, employed for the highly precise determination of dimensions.

MRI: Magnetic Resonance Imaging—a means of examining the invisible internals of items—usually human parts and organs, though the technique can be applied to non-living entities—with the combined use of powerful magnetic fields and high-frequency radio waves.

ORRERY: A usually clockwork mechanism which simulates, for entertainment or study, the motion of the celestial bodies around the sun or Earth.

PANTOGRAPH: A mechanism composed of jointed metal parallelograms which permits the exact copying of plans, diagrams or solid objects, since tracing the outline by one end of the device produces exactly similar movement of pens or cutting implements at the other.

PERMIAN: A 50-million-year-long geological period, beginning about 290 million years ago, and following the Carboniferous era. Often large thickness of sandstone or salts were laid down to form the capstone to oil and gas deposits from earlier times.

PHOTOLITHOGRAPHY: A form of printing in which a photographic image is transferred to a printing surface; today used in the manufacture of semiconductors.

PLANCK CONSTANT: Named for Max Planck, the German physicist, it relates the energy of a quantum of electromagnetic radiation to its frequency. It is usually represented by the symbol h.

PLASMA: A gas, usually formed at very high temperature, characterized by the presence of free electrons and positive ions.

PRECISION: Though often synonymous with accuracy and exactness, it is taken by engineers to mean the degree of refinement in a specification, the greater the number of zeroes after the decimal point suggesting the greater the precision of a measurement.

QUANTUM MECHANICS: The branch of physics concerned with the interaction and duality of atomic and subatomic particles and phenomena.

RATE (OF WARSHIPS): The classification of warships, in days of sail, according (by and large) to the number of guns: a first-rate ship would carry 100 cannon, a second rate 80, and so on.

REMONTOIR: The part of a clock that, by a system of weights and springs, keeps a constant power supply to the balance.

RUNAWAY: If an engine governor fails it is probable that the engine will run away, unchecked—which can be exceptionally dangerous.

SANSCULOTTE: So named—"without trousers"—because of their usual shabbiness, the term refers to the revolutionary poor in France who did much of the dirty work during the "Terror."

SEMICONDUCTOR: A material—silicon, germanium— that has conductivity which can be altered. It forms the basis of almost all transistors in small electronics.

SEXTANT, OCTANT: A mariner's hand-held instrument useful for employing the stars and planets for navigation. GPS has almost entirely supplanted the use of the sextant and its kin, the octant, though navigators still have to know how to use them.

SI: Only Burma, Liberia, and the United States have declined to sign up for the International System of Units, signified by the initials SI.

SILICON: One of the most common molecular constituents of the rocks that make up this planet, and a component now at the heart of most computer chips.

SLIDE REST: That part of a lathe which holds the various tools that can be brought to bear on a workpiece.

SLIDE RULE: Until recently all engineers carried a top-pocket slide rule, a portable calculating device of great speed and limited accuracy, which employed sliding rules marked with logarithmic scales, and the movement of which could give instant answers to arithmetical questions.

SPRINGS (BOATMANSHIP): The ropes which are used to secure a vessel to the dockside were called springs; the order to ease springs meant to loosen the ropes in preparation for sailing.

SPUD (OIL INDUSTRY): To begin the drilling of an oil well the bit is bounced up and down against the surface until a foot-deep hole has been created, after which drilling proper can begin.

TALLEYRAND: An eighteenth-century French diplomat and bishop—and prince—whose cunning, craftiness, and hauteur became a byword.

TAPPET: A small piece within a machine that rises and falls on contact with a camshaft, and can in doing so impart motion to a larger piece, such as a valve.

TEST MASS: The name given to the cylinder of fused silica mounted, along with a complex set of compensatory weights and springs, at the end of each chamber of the LIGO observatories.

TETSUBIN: A Japanese cast-iron kettle, with lid and handle, usually heated over a charcoal fire for making tea.

TOISE: Until it was phased out in the early nineteenth century, the toise was a measure of length in France, equivalent to about 1.9 meters. It was divided into 6 *pieds*.

TOLERANCE: The permissible variation in size from a specified standard allowed for a machined part. High precision demands high tolerances.

TRACEABILITY: The term refers to a chain by which any measurement may be traced back to a standard—so that, for example, the second on a wristwatch may be connected to and ultimately measured against the signal of constant time put out by an atomic clock.

TREMBLER COIL: An early spark generator used in the ignition systems of cars, such as the Model T Ford.

VANADIUM: The addition of this heavy silvery-gray element to steel greatly increases its hardness, and so forms the additive basis for many sophisticated alloys.

VERNIER: The sixteenth-century French mathematician who first created the scale that bears his name, and which allows for the measurement of fractions of the graduations on the main scale, against which a vernier scale is designed to slide.

VESPERS: Sunset prayers in a monastery, coming close to the end of the liturgical day.

WABI-SABI: The polar opposite of perfection; the aesthetic represented by this currently much-favored term denotes the appreciation of impermanence, roughness, and the quiet liking for craft.

Bibliography

Ackermann, Silke. *Director's Choice: Museum of the History of Science.* Oxford. Scala Arts & Heritage Publishers. 2016.

Adams, William Howard. *The Paris Years of Thomas Jefferson.* New Haven, CT. Yale University Press. 1997.

Albrecht, Albert B. *The American Machine Tool Industry: Its History, Growth and Decline—A Personal Perspective.* Richmond, IN. Privately published. 2009.

Alder, Ken. *The Measure of All Things: The Seven-Year Odyssey and Hidden Error that Transformed the World.* Boston. Little, Brown. 2002.

Allen, Lewis, et al. *The Hubble Space Telescope Optical*

Systems Failure Report. Washington, DC. NASA. 1990.

Althin, Torsten K. W. *C. E. Johansson 1864–1943: The Master of Measurement.* Stockholm. Aktiebolaget C. E. Johansson. 1948.

Atkins, Tony and Marcel Escudier. *Oxford Dictionary of Mechanical Engineering.* Oxford. Oxford University Press. 2013.

Atkinson, Norman. *Sir Joseph Whitworth: 'The World's Best Mechanician.'* Stroud, UK. Sutton Publishing. 1996.

Australian Transport Safety Bureau. *In-Flight Uncontained Engine Failure Overhead Batam Island, Indonesia. 4 November 2010.* Canberra. Australian Government. 2013.

Baggott, Jim. *The Quantum Story: A History in Forty Moments.* Oxford. Oxford University Press. 2011.

Baillie, G. H., C. Clutton, and C. A. Ilbert, *Britten's Old Clocks and Watches and Their Makers* (7th edition). New York. Bonanza Books. 1956.

Barnett, Jo Ellen. *Time's Pendulum.* New York. Plenum Press. 1998.

Bennett, Martin. *Rolls-Royce: The History of the Car.* New York. Arco. 1974.

Betts, Jonathan. *Harrison.* London. The National Maritime Museum. 1993.

Borth, Christy. *Masters of Mass Production.* New York. Bobbs-Merrill Co. 1945.

Bostrom, Nick. *Superintelligence: Paths, Dangers, Strategies.* Oxford. Oxford University Press. 2014.

Brown, Henry T. *Five Hundred and Seven Mechanical Movements.* New York. Brown and Seward. 1903.

Brown & Sharpe Mfg. Co. *Practical Treatise on Milling and Milling Machines.* Providence, RI. Brown & Sharpe. 1914.

Bryant, John and Chris Sangwin. *How Round Is Your Circle? Where Engineering and Mathematics Meet.* Princeton, NJ. Princeton University Press. 2008.

Burdick, Alan. *Why Time Flies: A Mostly Scientific Investigation.* New York. Simon & Schuster. 2017.

Cantrell, John and Gillian Cookson. *Henry Maudslay & the Pioneers of the Machine Age.* Stroud, UK. Tempus. 2002.

Carbone, Gerald M. *Brown and Sharpe and the Measure of American Industry.* Jefferson, NC. McFarland & Co. 2017.

Carey, Geo. G. *The Artisan; or Mechanic's Instructor.* London. William Cole. 1833.

CERN. *Infinitely CERN: Memories from Fifty Years of Research.* Geneva. Editions Suzanne Hurter. 2004.

Chandler, Alfred D. *Inventing the Electronic Century: The Epic Story of Consumer Electronics and Com-*

puter Industries. Cambridge, MA. Harvard University Press. 2005.

Chrysler, Walter P. *Life of an American Workman.* New York. Dodd, Mead & Co. 1937.

Collins, Harry. *Gravity's Kiss: The Detection of Gravitational Waves.* Cambridge, MA. MIT Press. 2017.

Cossons, Neil, Andrew Nahum, and Peter Turvey. *Making of the Modern World: Milestones of Science and Technology.* London. John Murray. 1992.

Crease, Robert P. *World in the Balance: The Historic Quest for an Absolute System of Measurement.* New York. W. W. Norton. 2011.

Crump, Thomas. *The Age of Steam: The Power that Drove the Industrial Revolution.* New York. Carroll & Graf. 2007.

Darrigol, Olivier. *A History of Optics: From Greek Antiquity to the Nineteenth Century.* Oxford. Oxford University Press. 2012.

Dawson, Frank. *John Wilkinson: King of the Ironmasters.* Stroud, UK. The History Press. 2012.

Day, Lance and Ian McNeil (eds). *Biographical Dictionary of the History of Technology.* London. Routledge. 1996.

Derry, T. K. and Trevor Williams. *A Short History of*

Technology: From the Earliest Times to AD 1900. Oxford. Oxford University Press. 1960.

DeVorkin, David and Robert W. Smith. *Hubble: Imaging Space and Time.* Washington, DC. National Geographic Society. 2008.

Dickinson, H. W. *John Wilkinson, Ironmaster.* Ulverston, UK. Hume Kitchin. 1914.

———. *Matthew Boulton.* Cambridge, UK. Cambridge University Press. 1937.

Duncan, David Ewing. *Calendar: Humanity's Epic Struggle to Determine a True and Accurate Year.* New York. Avon Books. 1998.

Dvorak, John. *Mask of the Sun: The Science, History and Forgotten Lore of Eclipses.* New York. Pegasus Books. 2017.

Easton, Richard D. and Eric F. Frazier. *GPS Declassified: From Smart Bombs to Smartphones.* Lincoln. University of Nebraska Press. 2013.

Evans, Chris. *Precision Engineering: An Evolutionary View.* Bedford, UK. Cranfield Press. 1989.

Fenna, Donald. *Dictionary of Weights, Measures and Units.* Oxford. Oxford University Press. 2002.

Free, Dan. *Early Japanese Railways 1853–1914: Engineering Triumphs that Transformed Meiji-Era Japan.* Rutland, VT. Tuttle Publishing. 2008.

Gleick, James. *Chaos: Making a New Science*. New York. Viking. 1987.

Golley, John. *Whittle: The True Story*. Shrewsbury, UK. Airlife Publishing Ltd. 1987.

Gordon, J. E. *Structures: or, Why Things Don't Fall Down*. London. Penguin. 1978.

Gould, Rupert T. *The Marine Chronometer: Its History and Development*. London. J. D. Potter. 1923.

Guye, Samuel. *Time & Space: Measuring Instruments from the 15th to the 19th Century*. New York. Praeger Publishers. 1971.

Hand, David J. *Measurement: A Very Short Introduction*. Oxford. Oxford University Press. 2016.

Hart-Davis, Adam (ed). *Engineers: From the Great Pyramids to the Pioneers of Space Travel*. New York. Dorling Kindersley. 2012.

Heffernan, Virginia. *Magic and Loss: The Internet as Art*. New York. Simon & Schuster. 2016.

Hiltzik, Michael. *Big Science: Ernest Lawrence and the Invention that Launched the Military-Industrial Complex*. New York. Simon & Schuster. 2015.

Hindle, Brooke. *Technology in Early America*. Chapel Hill. University of North Carolina Press. 1966.

Hindle, Brooke and Steven Lubar. *Engines of Change: The American Industrial Revolution, 1790–1860*. Washington, DC. Smithsonian. 1986.

Hirshfeld, Alan W. *Parallax: The Race to Measure the Cosmos.* New York. Henry Holt. 2002.

Hooker, Stanley. *Not Much of an Engineer: An Autobiography.* Shrewsbury, UK. Airlife Publishing. 1984.

Hounshell, David A. *From the American System to Mass Production, 1800–1932.* Baltimore. Johns Hopkins University Press. 1984.

Hunt, Robert (ed.). *Hunt's Hand-Book to the Official Catalogues: An Explanatory Guide...to the Great Exhibition.* London. Spicer Bros. 1851.

Johnson, George. *The Ten Most Beautiful Experiments.* New York. Knopf. 2008.

Johnson, Steven. *Where Good Ideas Come From: The Natural History of Innovation.* New York. Penguin. 2010.

Jones, Alexander. *A Portable Cosmos: Revealing the Antikythera Mechanism, Scientific Wonder of the Ancient World.* Oxford. Oxford University Press. 2017.

Jones, Tony. *Splitting the Second: The Story of Atomic Time.* Bristol, UK. Institute of Physics Publishing. 2000.

Kaempffert, Waldemar (ed.). *A Popular History of American Invention.* New York. A. L. Burt Company. 1924.

Kaplan, Margaret L., et al. *Precisionism in America*

1915–1941: Reordering Reality. New York. Harry N. Abrams. 1994.

Kaye, G. W. C. and T. H. Laby *Tables of Physical and Chemical Constants.* London. Longman. 1911. 13th edition. 1966.

Kirby, Ed. *Industrial Sharon: Sharon, Connecticut, in the Salisbury Iron District.* Sharon, CT. Sharon Historical Society. 2015.

Kirby, Richard Shelton, et al. *Engineering in History.* New York. McGraw-Hill. 1956.

Klein, Herbert Arthur. *The Science of Measurement: A Historical Survey.* New York. Dover Publications. 1974.

Kula, Witold. *Measures and Men.* Princeton, NJ. Princeton University Press. 1986.

Lacey, Robert. *Ford: The Men and the Machine.* Boston. Little, Brown. 1986.

Lager, James L. *Leica: An Illustrated History* (3 vols). Closter, NJ. Lager Limited Editions. 1993.

Leapman, Michael. *The World for a Shilling: How the Great Exhibition of 1851 Shaped a Nation.* London. Hodder Headline. 2001.

Lynch, Jack. *You Could Look It Up: The Reference Shelf from Ancient Babylon to Wikipedia.* London. Bloomsbury. 2016.

Madou, Marc J. *Fundamentals of Microfabrication:*

The Science of Miniaturization. Boca Raton, FL. CRC Press. 2002.

McNeil, Ian. *Joseph Bramah: A Century of Invention, 1749–1851.* Newton Abbot, UK. David & Charles. 1968.

Milner, Greg. *Pinpoint: How GPS Is Changing Technology, Culture, and Our Minds.* New York. W. W. Norton. 2016.

Mitutoyo Corporation. *A Brief History of the Micrometer.* Tokyo. 2011.

Moore, Wayne R. *Foundations of Mechanical Accuracy.* Bridgeport, CT. Moore Special Tool Co. 1970.

Muir, Diana. *Reflections in Bullough's Pond: Economy and Ecosystem in New England.* Lebanon, NH. University Press of New England. 2000.

Mumford, Lewis. *The Myth of the Machine: Technics and Human Development.* New York. Harcourt Brace. 1966.

Nahum, Andrew. *Frank Whittle: Invention of the Jet.* Cambridge, UK. Icon Books. 2005.

Noble, David F. *America by Design: Science, Technology, and the Rise of Corporate Capitalism.* New York. Knopf. 1977.

———. *Forces of Production: A Social History of Industrial Automation.* New York. Knopf. 1984.

———. *The Religion of Technology: The Divinity of*

Man and the Spirit of Invention. New York. Knopf. 1997.

Pearsall, Ronald. *Collecting and Restoring Scientific Instruments*. New York. Arco. 1974.

Penrose, Roger. *The Road to Reality: A Complete Guide to the Laws of the Universe*. London. Random House. 2004.

Pugh, Peter. *The Magic of a Name: The Rolls-Royce Story—The First Forty Years*. London. Icon Books. 2000.

Quinn, Terry. *From Artifacts to Atoms: The BIPM and the Search for Ultimate Measurement Standards*. New York. Oxford University Press. 2012.

Rid, Thomas. *Rise of the Machines: A Cybernetic History*. New York. W. W. Norton. 2016.

Rolls-Royce PLC. *The Jet Engine*. Chichester, UK. Wiley. 2005.

Rolt, L. T. C. *Tools for the Job: A Short History of Machine Tools*. London. Batsford. 1965.

Rosen, William. *The Most Powerful Idea in the World: A Story of Steam, Industry and Invention*. Chicago. University of Chicago Press. 2010.

Roser, Christoph. *"Faster, Better, Cheaper" in the History of Manufacturing: From the Stone Age to Lean Manufacturing and Beyond*. Boca Raton, FL. CRC Press. 2017.

Russell, Ben. *James Watt: Making the World Anew.* London. Reaktion Books. 2014.

Rybczynski, Witold. *One Good Turn: A Natural History of the Screwdriver and the Screw.* New York. Touchstone/Simon & Schuster. 2000.

Schivelbusch, Wolfgang. *The Railway Journey: The Industrialization of Time and Space in the Nineteenth Century.* Oakland. University of California Press. 1977.

Schlosser, Eric. *Command and Control: Nuclear Weapons, the Damascus Incident, and the Illusion of Safety.* New York. Penguin. 2013.

Setright, L. J. K. *Drive On! A Social History of the Motor Car.* London. Granta Books. 2003.

Singer, Charles, et al. *A History of Technology: Vol IV: The Industrial Revolution, 1750–1850.* Oxford. Oxford University Press. 1958.

Smil, Vaclav. *Prime Movers of Globalization: The History and Impact of Diesel Engines and Gas Turbines.* Cambridge, MA. MIT Press. 2010.

Smiles, Samuel. *Lives of the Engineers* (5 vols). London. 1862.

———. *Industrial Biography: Iron Workers and Tool Makers.* London. 1863.

Smith, Gar. *Nuclear Roulette: The Truth about the Most Dangerous Energy Source on Earth.* White

River Junction, VT. Chelsea Green Publishing. 2012.

Smith, Merritt Roe. *Harpers Ferry Armory and the New Technology: The Challenge of Change*. Ithaca, NY. Cornell University Press. 1977.

Sobel, Dava. *Longitude: The True Story of a Lone Genius Who Solved the Greatest Scientific Problem of His Time*. New York. Walker and Company. 1995.

Soemers, Herman. *Design Principles for Precision Mechanisms*. Eindhoven, Netherlands. 2011.

Standage, Tom. *The Turk: The Life and Times of the Famous Eighteenth-Century Chess Playing Machine*. New York. Walker and Company. 2002.

Stephens-Adamson Manufacturing Company. *General Catalog No. 55*. Aurora, IL. Stephens-Adamson Mfg. Co. 1941.

Stoddard, Brooke C. *Steel: From Mine to Mill, the Metal that Made America*. Minneapolis. Quarto. 2015.

Stout, K. J. *From Cubit to Nanometre: A History of Precision Measurement*. Teddington, UK. National Physical Laboratory. Penton Press. No date.

Tomlinson, Charles. *Cyclopaedia of Useful Arts, Mechanical and Chemical, Manufactures, Mining, and Engineering* (3 vols.). London. Virtue and Company. 1866.

Tsujimoto, Karen. *Images of America: Precisionist Painting and Modern Photography.* San Francisco. San Francisco Museum of Modern Art. 1982.

Usher, Abbott Payson. *A History of Mechanical Inventions.* Cambridge, MA. Harvard University Press. 1929.

Utterback, James M. *Mastering the Dynamics of Innovation.* Boston. Harvard Business School Press. 1994.

Vessey, Alan. *By Precision into Power: A Bicentennial Record of D. Napier & Son.* Stroud, UK. Tempus. 2007.

Wagner, Erica. *Chief Engineer: Washington Roebling: The Man Who Built the Brooklyn Bridge.* New York. Bloomsbury. 2017.

Watson, Peter. *Ideas: A History of Thought and Invention, from Fire to Freud.* New York. Harper. 2005.

Wilczek, Frank. *A Beautiful Question: Finding Nature's Deep Design.* New York. Viking. 2015.

Wise, M. Norton (ed.). *The Values of Precision.* Princeton, NJ. Princeton University Press. 1995.

Wolfram, Stephen. *Idea Makers: Personal Perspectives on the Lives and Ideas of Some Notable People.* London. Wolfram LLC. 2016.

Yapp, G. W. (compiler). *Official Catalogue of the Great*

Exhibition of the Works of Industry of All Nations 1851. London. Spicer Bros. 1851.

Zimmerman, Robert. *The Universe in a Mirror: The Saga of the Hubble Telescope and the Visionaries Who Built It.* Princeton, NJ. Princeton University Press. 2008.

About the Author

SIMON WINCHESTER is the acclaimed author of many books, including *The Professor and the Madman*, *The Men Who United the States*, *Atlantic*, *Pacific*, *The Man Who Loved China*, *A Crack in the Edge of the World*, and *Krakatoa*, all of which were *New York Times* bestsellers and appeared on numerous best-of and notable lists. In 2006, Mr. Winchester was made an officer of the Order of the British Empire (OBE) by Her Majesty the Queen. He lives in western Massachusetts.

www.simonwinchester.com

THE NEW LUXURY IN READING

We hope you enjoyed reading
our new, comfortable print size and found it
an experience you would like to repeat.

Well – you're in luck!

HarperLuxe offers the finest in fiction and
nonfiction books in this same larger print size and
paperback format. Light and easy to read, HarperLuxe
paperbacks are for book lovers who want to see
what they are reading without the strain.

For a full listing of titles and
new releases to come, please visit our website:

www.HarperLuxe.com